CW01360053

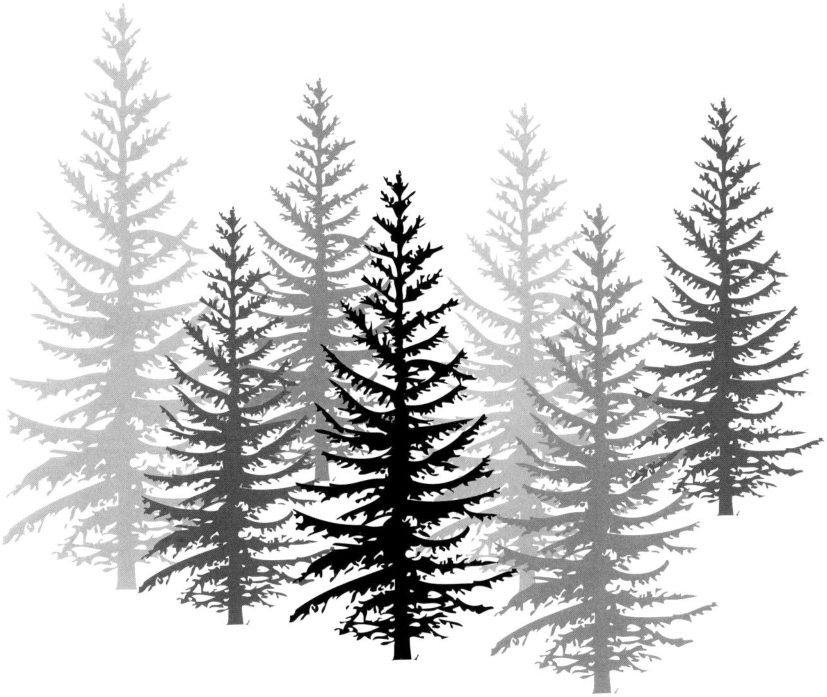

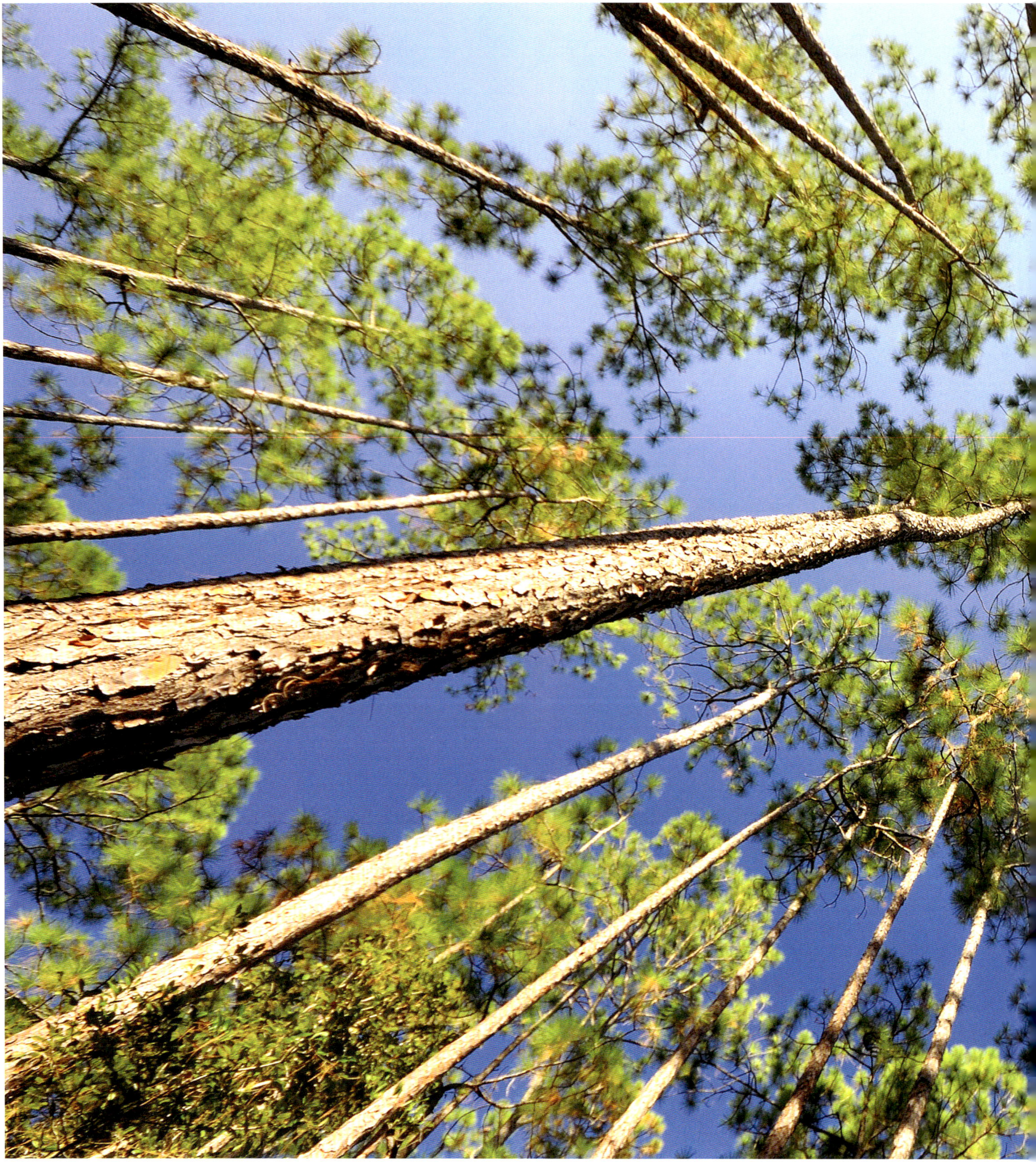

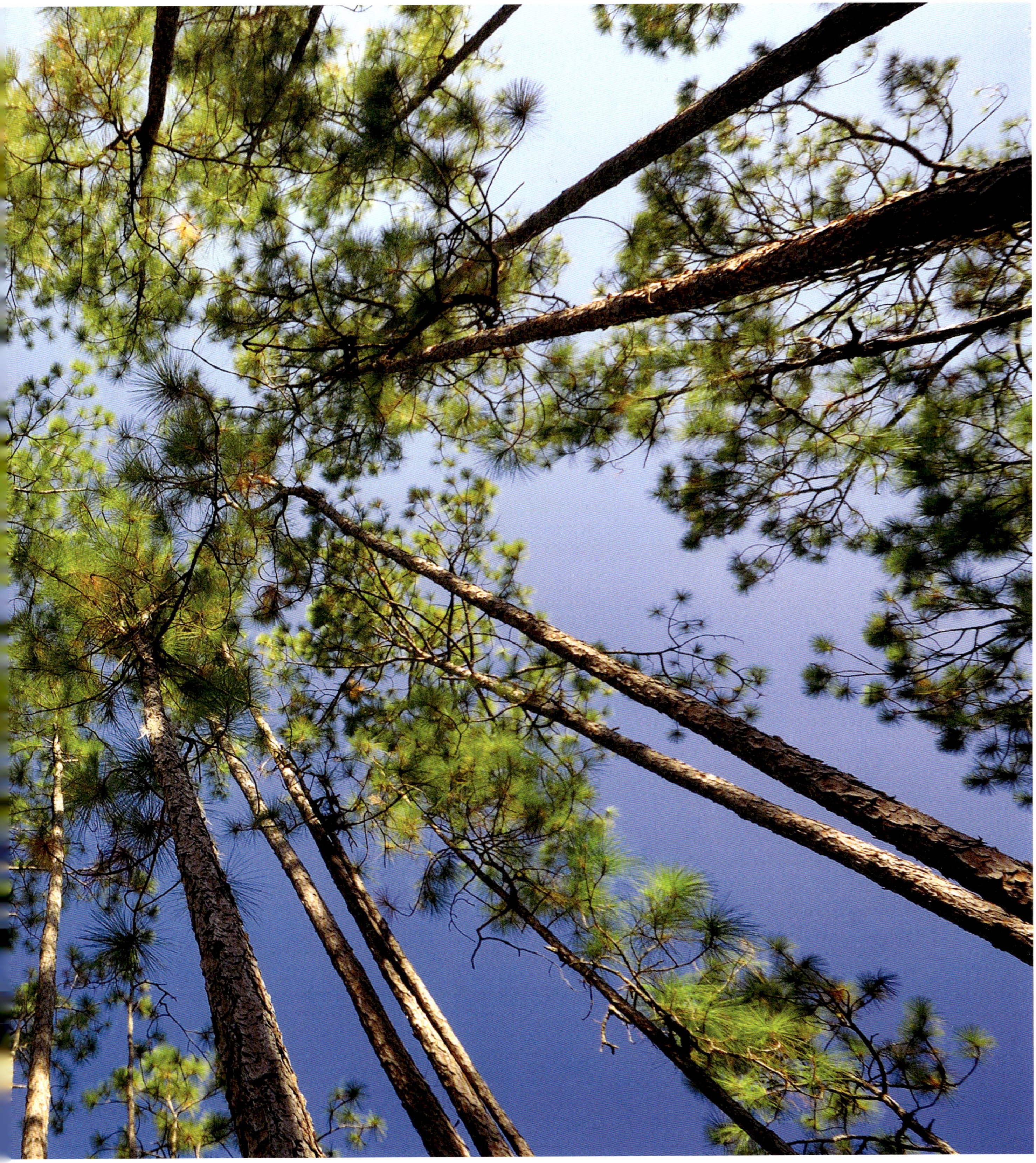

LOFT

2016 © LOFT Publications, S.L.
c/ Domènech, 7-9, 2º 1ª
08012 Barcelona, Spain
T: +34 93 268 80 88

ISBN 978-84-9936-972-3 (EN)

booq publishing

2016 © booq publishing, S.L.
c/ València, 93, Pral. 1a
08029 Barcelona, Spain
www.booqpublishing.com

ISBN 978-84-944830-11 (DE)

Editorial coordination: Claudia Martínez Alonso
Art direction: Mireia Casanovas Soley
Edition: Francesc Zamora
Graphic edition: Manel Gutiérrez (@mgutico)
Texts: Francesc Zamora, Yuri Caravaca
Layout: Cristina Simó, Sara Abril, David Andreu
Translations: Textcase

Printed in Spain

LOFT affirms that it possesses all the necessary rights for the publication of this material and has duly paid all royalties related to the authors' and photographers' rights. **LOFT** also affirms that is has violated no property rights and has respected common law, all authors' rights and other rights that could be relevant. Finally, **LOFT** affirms that this book contains neither obscene nor slanderous material.

The total or partial reproduction of this book without the authorization of the publishers violates the two rights reserved; any use must be requested in advance.

In some cases it might have been impossible to locate copyright owners of the images published in this book. Please contact the publisher if you are the copyright owner in such a case.

6	Introduction	70	*Larix laricina*	134	*Mangifera indica*	194	*Chorisia speciosa*
10	Xiloteca Manuel Soler	72	*Pinus rigida*	136	*Melia azedarach*	196	*Tilia americana*
		74	*Pinus sylvestris*	138	*Acer campestre*	198	*Tilia vulgaris*
14	*Cryptomeria japonica*	76	*Chamaecyparis lawsoniana*	140	*Swietenia mahagoni*	200	*Heritiera littoralis*
16	*Pinus patula*	78	*Pinus echinata*	142	*Carapa guianensis*	202	*Adansonia digitata*
18	*Podocarpus totara*	80	*Pinus taeda*	144	*Khaya senegalensis*	204	*Cavanillesia platanifolia*
20	*Thuja occidentalis*	82	*Pseudotsuga menziesii*	146	*Acer palmatum*	206	*Hibiscus tiliaceus*
22	*Abies lasiocarpa*	86	*Podocarpus gracilior*	148	*Acer saccharum*	208	*Thespesia populnea*
24	*Thuja plicata*	88	*Pinus elliottii*	150	*Calodendrum capense*	210	*Bassia latifolia*
26	*Cupressus lusitanica*	90	*Araucaria angustifolia*	152	*Ailanthus altissima*	212	*Schima wallichii*
28	*Picea abies*	92	*Agathis robusta*	154	*Cedrela fissilis*	214	*Arbutus menziesii*
30	*Pinus lambertiana*	94	*Dacrydium cupressinum*	156	*Harpephyllum caffrum*	216	*Diospyros kaki*
32	*Pinus strobus*	96	*Agathis australis*	158	*Rhus typhina*	218	*Diospyros virginiana*
34	*Picea engelmannii*	98	*Cedrus libani*	160	*Sapindus saponaria*	220	*Cariniana legalis*
36	*Abies balsamea*	100	*Pinus radiata*	162	*Toona ciliata*	222	*Manilkara zapota*
38	*Pinus monticola*	102	*Pinus contorta*	164	*Prunus serotina*	224	*Sassafras albidum*
40	*Pinus ponderosa*	104	*Pinus palustris*	166	*Celtis occidentalis*	226	*Cinnamomum camphora*
42	*Calocedrus decurrens*	106	*Juniperus virginiana*	168	*Zelkova serrata*	228	*Ocotea porosa*
44	*Abies concolor*	108	*Pinus caribaea*	170	*Prunus avium*	230	*Persea americana*
46	*Sequoia sempervirens*	110	*Taxus baccata*	172	*Brosimum alicastrum*	232	*Platanus occidentalis*
48	*Tsuga canadensis*	112	*Agathis dammara*	174	*Pyrus communis*	236	*Grevillea robusta*
50	*Picea sitchensis*	114	*Bursera simaruba*	176	*Celtis africana*	238	*Platanus hybrida*
52	*Taxodium distichum*	116	*Aesculus glabra*	178	*Maclura pomifera*	240	*Faurea saligna*
56	*Picea mariana*	118	*Acer negundo*	180	*Artocarpus altilis*	242	*Nyssa sylvatica*
58	*Tsuga heterophylla*	120	*Aesculus hippocastanum*	182	*Celtis australis*	244	*Cornus florida*
60	*Cupressus sempervirens*	122	*Swietenia macrophylla*	184	*Celtis sinensis*	246	*Liquidambar styraciflua*
62	*Fitzroya cupressoides*	124	*Acer macrophyllum*	186	*Hemiptelea davidii*	248	*Cercidiphyllum japonicum*
64	*Pinus resinosa*	126	*Acer rubrum*	188	*Ziziphus jujuba*	250	*Balanites aegyptiaca*
66	*Pinus banksiana*	130	*Flindersia brayleyana*	190	*Ochroma pyramidale*	252	*Guaiacum officinale*
68	*Chamaecyparis nootkatensis*	132	*Sclerocarya birrea*	192	*Ceiba pentandra*	254	Credits and related websites

The benefits we derive from forest and their trees are widespread in various spheres: they influence the environment in which we live by moderating climate, improving air quality, maintaining high water tables, and harboring wildlife. Economically, their wood is a renewable resource that has been exploited by every human culture on every continent, prized for the ease with which it can be worked, its strength, versatility, and above all its beauty.

It is in our interest that we develop our knowledge about tree species and forest ecosystems, making sure that wood is produced in a sustainable manner and that overexploitation and international trade do not threaten their survival. Wood identification is important for the organization of the data especially in the context of the IUCN Red List of Threatened Species (www.iucnredlist.org; see Table 1, p. 8) and the Convention on International Trade in Endangered Species of Wild Fauna and Flora (www.cites.org).

Xylotomy deals with the comparative study of different woods and is an indispensable tool for understanding the scientific and economic value of this wonderful material. This field of study, which also focuses on industrial uses of wood, also informs our forest management plans. Secondly, it helps users choose the best wood for an application based on the physical and mechanical wood properties. In this lies the value of a collection of wood samples from everywhere in the world, which fulfills the function of a template allowing the identification of similarities with any given sample.

This volume presents descriptions of over 100 trees and bushes from all parts of the globe and their utility as a material. We have drawn on many sources in compiling tables of the physical and mechanical properties of the featured woods into a useful reference guide. The unique physical properties of each type of wood determines its suitability for different uses. These properties, including bending strength, density, hardness, stiffness and shrinkage rate must be considered to use wood to its best advantage. For instance, some species are suitable for boat building because they are durable and resistant to decay, while other species are extensively used in building construction because of their structural strengths. Texture grain and design determine the potential uses of the different woods and inform users of their response to tools under common woodworking procedures as described in a color coded chart (see Table 2, p. 8). Hardness is an important property varying for each species, and its measurement —in pounds-force— is essential to identify appropriate choices. Taking into account this valuable characteristic particularly in lumber and timber trade, the list of trees and bushes is organized from softest to hardest within each order of the biological classification. For instance, white sassafras, camphor tree, Brazilian walnut, and avocado, all in the Laurales Order category of the biological classification, are listed as white sassafras being the softest and avocado being the hardest; within the Pinales Order category, Japanese cedar is among the softest and yew being among the hardest; within the Sapindales Order category, Ohio buckeye is one of the softest woods and Australian red cedar being the hardest.

Der Nutzen, den wir aus Wald und Bäumen ziehen, ist vielfältig und kann nicht auf einen Bereich begrenzt werden: Die positive Beeinflussung des Klimas, die Verbesserung der Luftqualität, die Erhaltung eines hohen Grundwasserspiegels und die Beherbergung von Flora und Fauna. Wirtschaftlich betrachtet ist Holz ein regenerativer Rohstoff, der seit jeher von jeder menschlichen Kultur auf jedem Kontinent genutzt worden ist. Sein hoher Wert liegt in seiner einfachen Bearbeitbarkeit, seiner Härte, Vielseitigkeit und vor allem in seiner Schönheit begründet.

Um sicherzugehen, dass Holz in nachhaltiger Form produziert wird und Raubbau und internationaler Handel das Überleben des Waldes nicht gefährden, sollten wir in eigenem Interesse handeln und unser Wissen über Baumarten und das Ökosystem Wald erweitern. Besonders in Verbindung mit der Roten Liste gefährdeter Arten der Weltnaturschutzunion IUCN (IUCN Red List of Threatened Species; www.iucnredlist.org; siehe Tabelle 1, s. 8) und dem Übereinkommen über den internationalen Handel mit gefährdeten Arten freilebender Tiere und Pflanzen (Convention on International Trade in Endangered Species of Wild Fauna and Flora; www.cites.org) ist die Baumartenbestimmung wichtiger Bestandteil der Organisation von Daten.

Die Xylotomie beschäftigt sich mit der vergleichenden Untersuchung verschiedener Holzarten und ist unabdingbares Instrument, wenn es darum geht, den wissenschaftlichen und wirtschaftlichen Wert dieses wunderbaren Materials zu erfassen. Dieser Forschungsbereich, der sich ebenfalls mit der industriellen Verwendung von Holz auseinandersetzt, ist zudem Informationsquelle für unsere Waldwirtschaftsplanung. Er hilft uns außerdem bei der Auswahl der am besten für eine bestimmte Verwendung geeigneten Holzart, basierend auf deren Beschaffenheit und mechanischen Eigenschaften. Und genau darin liegt der Wert einer Sammlung von Holzproben aus allen Teilen der Welt. Sie übernehmen Vergleichsfunktion und ermöglichen so anhand eines beliebigen Musters die Identifikation von Ähnlichkeiten.

Dieser Buchband enthält Beschreibungen von über 100 Baum- und Buscharten aus allen Teilen der Welt und deren Nutzbarkeit als Material. Zur Erstellung dieses praktischen Nachschlagewerks haben wir eine Vielzahl verschiedener Quellen konsultiert, um die physischen und mechanischen Eigenschaften der aufgeführten Holzarten zusammenzutragen und tabellarisch darzustellen. Für welche Art der Verwendung eine Holzart geeignet ist bestimmen seine einzelnen physischen Eigenschaften. Diese Eigenschaften, wie Biegefestigkeit, Dichte, Härte, Festigkeit und Schwindungsrate müssen berücksichtigt werden, um eine Holzart bestmöglich einsetzen zu können. Manche Arten werden beispielsweise bevorzugt im Bootsbau verwendet, weil sie langlebig sind und nicht so schnell faulen. Andere Arten hingegen werden aufgrund ihrer starken Struktur verstärkt in der Baukonstruktion eingesetzt. Wie im Diagramm dargestellt (siehe Tabelle 2, s. 8), bestimmen Textur, Maserung und Design die potentiellen Einsatzmöglichkeiten verschiedenen Holzarten und informieren Verbraucher über das Verhalten des Holzes bei der Anwendung verschiedener Werkzeuge und gebräuchlicher Bearbeitungsmethoden. Die Härte ist eine entscheidende Eigenschaft, die von Holzart zu Holzart variiert. Sie wird in Pounds-force gemessen und ist von entscheidender Bedeutung, wenn es darum geht, die richtige Wahl zu treffen. Die Berücksichtigung dieser wertvollen Eigenschaft ist, besonders was den Handel mit Schnitt- und Bauholz anbelangt, von großer Bedeutung. Die Liste der Bäume und Büsche ist innerhalb jeder biologischen Klassifizierungsfolge von weich nach hart sortiert aufgeführt worden. Sassafrasbaum, Kampferbaum, Brasilianische Walnuss, und Avocado beispielsweise gehören bei der biologischen Klassifizierung allesamt zur Kategorie Laurales. Innerhalb dieser Kategorie ist der Sassafrasbaum als weichster und der Avocadobaum als härtester gelistet; innerhalb der Kategorie der Koniferen gehört die Japanische Zeder zu den weichsten und die Eibe zu den härtesten; innerhalb der Kategorie Sapindales ist die Ohio-Rosskastanie eines der weichsten Hölzer und die Australische Rotzeder eines der härtesten.

Classification based on the IUCN Red List of Threatened Species	
Extinct	EX
Extinct in the Wild	EW
Critically Endangered	CR
Endangered	EN
Vulnerable	VU
Conservation dependent	CD
Near Threatened	NT
Least Concern	LC
Data Deficient	DD
Not Evaluated	NE*
* Not yet assessed for the IUCN as of 5 September 2012)	

Table 1

Legend machinability	
	Poor to very poor results
	Poor results
	Moderate
	Fair to good results
	Very good to excellent results

Table 2

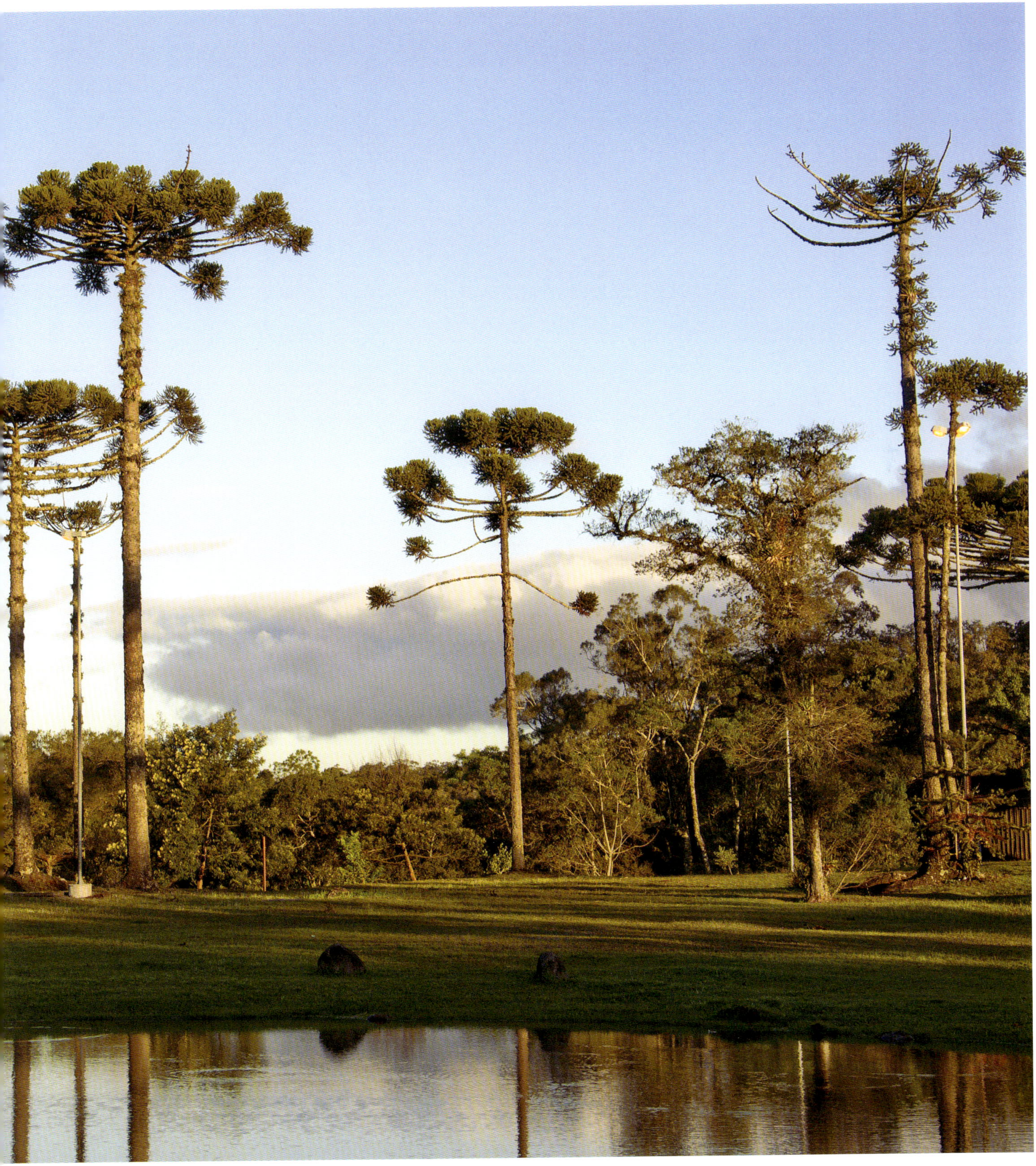

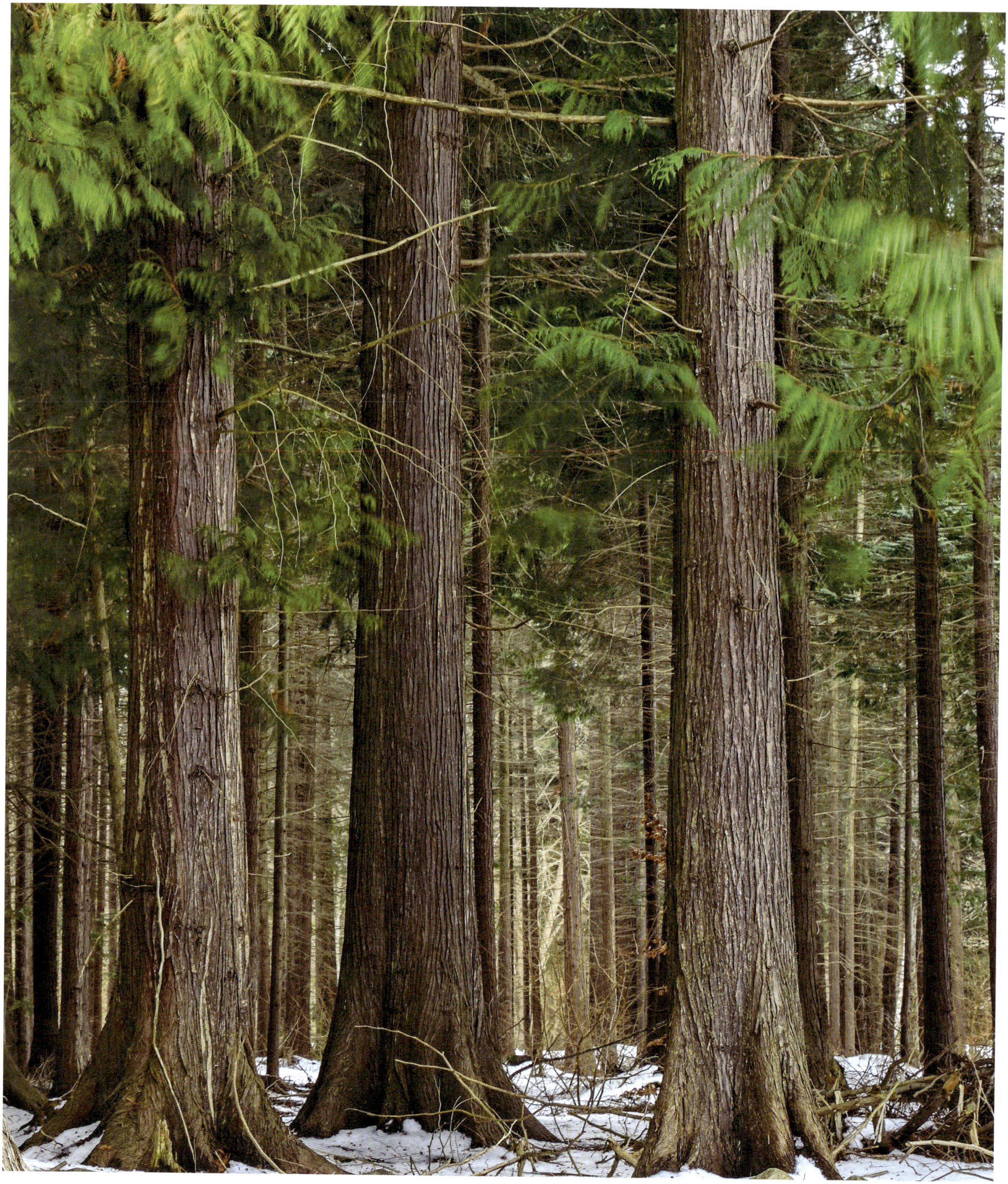

Xiloteca Manuel Soler

Mil maderas, published by the Polytechnic University of Valencia, Spain, 2001

Mil maderas II, published by the Polytechnic University of Valencia, Spain, 2004

Mil maderas III, published by the Polytechnic University of Valencia, Spain, 2008

While *Mil maderas IV* is waiting to be published, Mr. Soler is currently working on the V volume of the *Mil maderas* series

Interior shot of the Xiloteca Manuel Soler in Denia, Spain

Manuel Soler working at his desk

Manuel Soler

This book has in part been possible thanks to the generous collaboration of a wood collector based in Denia, Spain. With about 4,000 wood samples from all over the world, the Xiloteca Manuel Soler is one of the largest xylotheques in Europe. It is displayed in a small house surrounded by trees and built with Swedish pine by the owner himself.

Despite the little support that wood collectors receive in general from their respective governments, the Xiloteca Manuel Soler is open to researchers and general public. Inside, the walls are lined with wood samples perfectly identified and a good number of books. This is where Manuel Soler spends most of his time. A former captain in the merchant marine, he has collected wood samples from South Africa, Costa Rica, Cuba, Australia, Indonesia, the Fiji Islands and the Philippines. But not only are wood blocks found in the house: bamboo flutes, eucalyptus boomerangs and pieces of dugout canoes brought from Nigeria are an indication of the diverse ways in which wood is utilized.

Manuel Soler's interest and extensive knowledge about the different types of wood have brought him to edit several books published by the Polytechnic University of Valencia.

Dieses Buch wurde zu einem Teil dank der großzügigen Mithilfe eines Holzsammlers ermöglicht, der im spanischen Denia ansässig ist. Mit etwa 4 000 Holzproben aus allen Teilen der Welt zählt die Xiloteca Manuel Soler zu den größten Holzsammlungen Europas. Ausgestellt wird das Holz in einem kleinen, von Bäumen umgebenen Haus, das der Besitzer selbst aus schwedischer Pinie gebaut hat.

Trotz der geringen Unterstützung, die Holzsammler in der Regel von Seiten der jeweiligen Regierung erhalten, ist die Xiloteca Manuel Soler für Forscher und die Allgemeinheit geöffnet. Neben einer Vielzahl von Büchern reihen sich an den Wänden in ihrem Inneren exakt bestimmte Holzproben aneinander. Hier verbringt Manuel Soler den größten Teil seiner Zeit. Als früherer Kapitän bei der Handelsmarine hat er Holzproben aus Südafrika, Costa Rica, Kuba, Australien, Indonesien, den Fidschi Inseln und den Philippinen zusammengetragen. Es finden sich jedoch nicht ausschließlich Holzproben in dem Haus: Bambusflöten, Eukalyptus-Boomerangs und Teile eines Einbaums, Mitbringsel aus Nigeria, lassen erkennen, wie vielfältig Holz verwendet werden kann.

Sein Interesse und großes Wissen über die verschiedenen Arten von Holz haben Manuel Soler dazu bewegt, mehrere Bücher herauszugeben, die durch die Polytechnische Universität von Valencia veröffentlicht wurden.

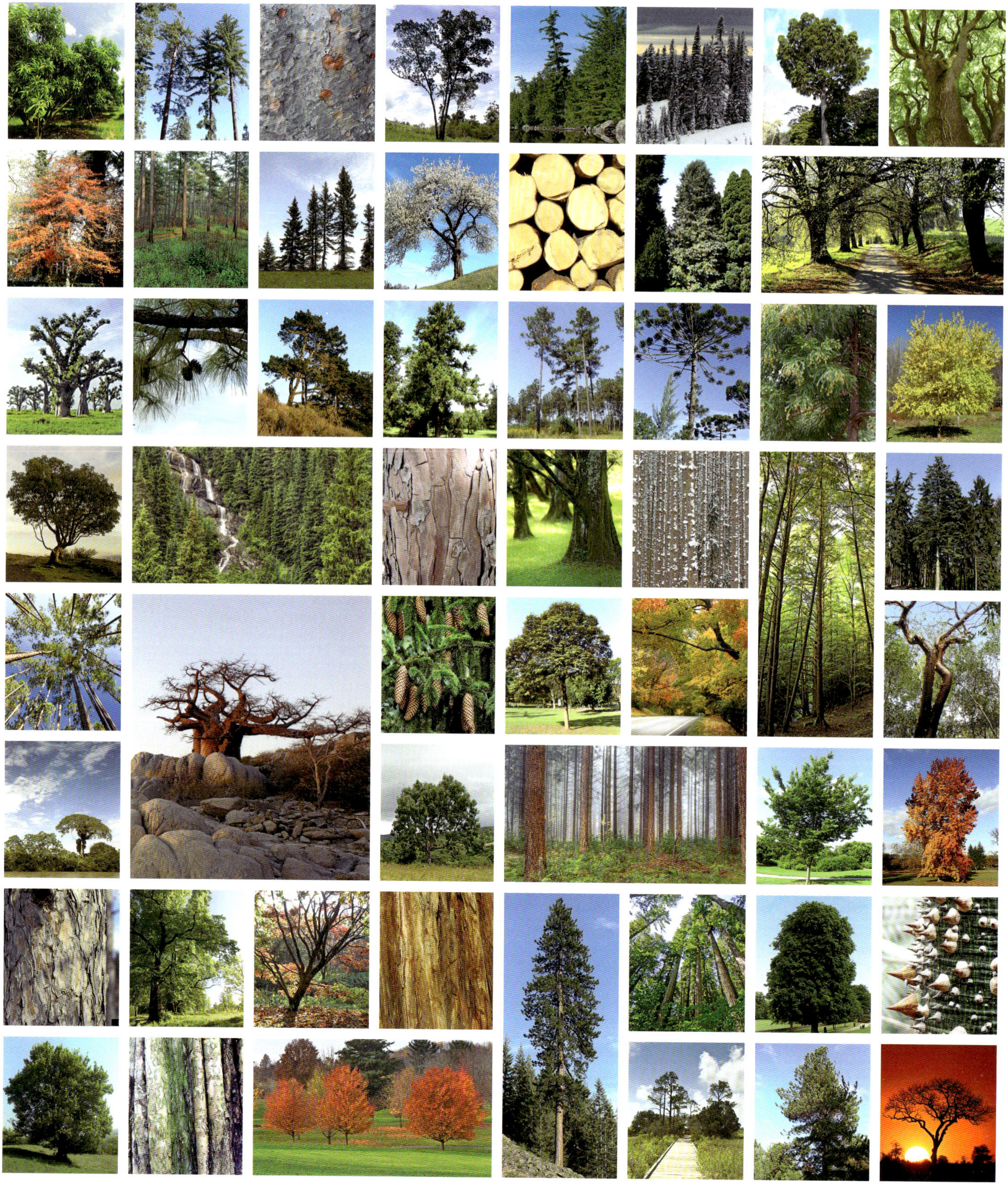

13

CRYPTOMERIA JAPONICA

Description

This is a genus of conifer native to Japan and consisting of a single species, also known in the Anglophone world as the Japanese cedar but known today as "sugi", which is the original Japanese name. Far from a cedar, it is distantly related to the giant sequoias, with which it shares its huge dimensions: up to 70 m (230 ft) tall, with diameters of up to 4 m (13 ft). The grain of its wood is typically straight, but it sometimes features irregular patterns in yellow or brown tones that contrast with the black vertical streaks of the resin. The sapwood is the color of cinnamon; the heartwood initially ranges from yellow to reddish in color, but oxidation turns it a pale brown.

Beschreibung

Die Sicheltanne gehört zur Gattung der Koniferen und ist als einzige ihrer Art in Japan heimisch. Früher in der angelsächsischen Welt als Japanische Zeder bekannt, ist heute, dem japanischen Original folgend, der Name „Sugi" gängig. Weit davon weg eine Zeder zu sein, ist sie entfernt mit den Sequoias, den Mammutbäumen, verwandt, mit denen sie ihre gigantische Größe teilt: bis zu 70 m Höhe und bis zu 4 m Stammdurchmesser. Die Holzmaserung ist typischerweise geradlinig angelegt, mit zufällig auftretenden gelben oder rotbraunen Mustern, die sich von den schwarzen vertikalen Adern des Harzes abheben. Das Splintholz ist zimtfarben, das Kernholz hat im Original eine gelbe bis rötliche Färbung, oxidiert aber zu einem blassen Braun.

Scientific / Botanical name
Cryptomeria japonica

Trade / Common name
Japanese cedar

Family name
Cupressaceae

Regions / Countries of distribution
Japan

Global threat status
NT

Common uses

Balustrades, packaging and crates, cabinetry, barrels, decorative plywood, construction materials, stairs, windows, flooring, furniture, etc.

Allgemeine Verwendung

Balustraden, Verpackung und Kisten, Tischlereiprodukte, Fässer, Sperrholz, Konstruktionsmaterial, Treppen, Fenster, Böden, Mobiliar, etc.

Machinability

Boring	
Nailing	
Painting	
Planing	
Polishing	
Screwing	
Staining	
Steam bending	

Physical properties

Numerical data	Green	Dry	English / Metric
Bending strength	3,409 // 239	5,115 // 359	psi // kgf/cm^2
Density		25 // 400	lbs/ft^3 // kg/m^3
Hardness		318 // 144	lbs // kg
Impact strength		16 // 40	in // cm
Maximum crushing strength	1,198 // 84	2,264 // 159	psi // kgf/cm^2
Shearing strength		884 // 62	psi // kgf/cm^2
Stiffness	1,020 // 71	1,200 // 84	1,000 psi // 1,000 kgf/cm^2
Weight	25 // 400	24 // 384	lbs/ft^3 // kg/m^3
Radial shrinkage		2	%
Tangential shrinkage		6	%

PINUS PATULA

Description

Patula pine is an evergreen conifer growing up to 30 m (98 ft) or higher and can attain a trunk diameter of 1.2 m (4 ft). It occasionally forks at the base to produce two straight and cylindrical trunks. When this happens the crown tends to spread to look round rather than spire-like. The bark is reddish and scaly when young, maturing gray-brown and vertically fissured. Patula pine bears long needles in droopy bundles of 2 to 5, and large cones borne on short stalks individually or in small clusters. They are reddish brown with a flat surface or slightly raised scales. Heartwood is light pink-brown, while sapwood is yellow-white, but there is not a demarcated distinction between the two.

Beschreibung

Pinus patula ist eine immergrüne Konifere, die 30 m oder höher wird und einen Stammdurchmesser von 1,2 m erreichen kann. In manchen Fällen gabelt er an der Basis, sodass zwei gerade und zylindrische Stämme entstehen. Wenn das geschieht, wird die Krone meist etwas breiter und sieht dann eher rund als spitz aus. In jungen Jahren sieht die Rinde schuppig aus und hat eine rötliche Farbgebung. Im Laufe der Jahre entstehen vertikale Spalten und die Farbe wechselt in ein Graubraun. Neben große Zapfen, die einzeln oder in kleinen Gruppen an kurzen Stielen wachsen, trägt Pinus Patula lange Nadeln, die in 2er bis 5er-Bündeln herabhängen. Die Zapfen sind rotbraun mit glatter Oberfläche oder leicht angehobenen Schuppen. Während das Kernholz ein helles Pink-Braun besitzt, ist das Splintholz Gelb-Weiß, wobei sich beide nicht nennenswert voneinander unterscheiden.

Scientific / Botanical name
Pinus patula

Trade / Common name
Patula pine

Family name
Pinaceae

Regions / Countries of distribution
Mexico

Global threat status
VU

Common uses
The uses of patula pine depend on the wood quality since knots are more or less numerous. Taking this into account end uses include boxes and crates, veneer, paneling, formwork, and particle boards.

Allgemeine Verwendung
Die Verwendbarkeit von *Pinus patula* hängt von der Holzqualität ab, da die Anzahl der Astlöcher recht hoch ausfallen kann. Zu seinen Verwendungszwecken gehören daher auch Kisten und Kästen, Furnier, Holzvertäfelung, Verschalung und Spanplatten.

Machinability

Boring	
Gluing	
Mortising	
Moulding	
Painting	
Planing	
Polishing	
Staining	

Physical properties

Numerical data	Green	Dry	English / Metric
Bending strength	5,465 // 384	8,490 // 596	psi // kg/cm^2
Density		29 // 464	lbs/ft^3 // kg/m^3
Hardness		320 // 145	lbs // kg
Impact strength		15 // 38	in // cm
Maximum crushing strength	3,025 // 212	5,160 // 362	psi // kg/cm^2
Shearing strength		1,170 // 82	psi // kg/cm^2
Stiffness	1,130 // 79	1,325 // 93	1,000 psi // 1,000 kg/cm^2
Weight	27 // 432	24 // 384	lbs/ft^3 // kg/m^3
Radial shrinkage		3	%
Tangential shrinkage		5	%

PODOCARPUS TOTARA

Description

The slow growing *Podocarpus totara* is an evergreen that thrives around river flats and sub-alpine forest elevations. It grows up to 30 m (98 ft) with a trunk diameter that can exceed 2 m (6.5 ft). The crown is dense in young specimens and opens as the tree ages. The bark, thick and furrowed, peels off in strips. The olive green leaves are short and prickly. The tōtara is a dioecious tree, which means that male and female flowers grow on different plants. Male flowers form yellow-green katkins. The female is a solitary pinkish small cone that grows into a fleshy red berry with a green seed. The heartwood of the tōtara is reddish brown and the sapwood a pale brown. Growth rings are very distinctive and the grain is straight, allowing it to be easily split.

Beschreibung

Podocarpus totara ist langsamwüchsig, immergrün und in der Umgebung von Flussebenen und auf sub-alpinen Hochebenen zu finden. Mit einem Stamm, der den Umfang von 2 m überschreiten kann, erlangt der Totara eine Höhe von bis zu 30 m. Während die Krone junger Exemplare dicht gewachsen ist, öffnet sie sich mit steigendem Alter. Die dicke, furchige Rinde schält sich in Streifen ab. Seine olivgrünen Blätter sind kurz und stachelig. Der Totara ist zweihäusig, was bedeutet, dass weibliche und männliche Blüten auf getrennten Individuen vorkommen. Männliche Blüten bilden gelbgrüne Kätzchen aus. Die weibliche Blüte ist ein einzelner kleiner Zapfen in Blassrosa, der zu einer fleischigen roten Beere mit grünem Samen heranwächst. Das Kernholz des Totara ist rotbraun, das Splintholz blassbraun. Die Wachstumsringe des Totara sind stark ausgeprägt und die gerade verlaufende Maserung trägt dazu bei, dass das Holz leicht gespalten werden kann.

Scientific / Botanical name
Podocarpus totara

Trade / Common name
Tōtara

Family name
Podocarpaceae

Regions / Countries of distribution
New Zealand

Global threat status
LC

Common uses
Prized by the Maori to fashion their war canoes, the timber of the totara is light in weight, easy to work and very durable. It is also used for cabinet making and carving.

Allgemeine Verwendung
Das Schnittholz ist von geringem Gewicht, einfach zu bearbeiten und langlebig, was es für die Maori zu einem wertvollen Bestandteil ihrer Kriegsboote machte. Es wird zudem für die Herstellung von Schränken und für Schnitzarbeiten verwendet.

Machinability

Boring	
Carving	
Gluing	
Mortising	
Moulding	
Nailing	
Planing	
Polishing	
Routing and recessing	
Sanding	
Screwing	
Turning	
Veneering qualities	

Physical properties

Numerical data	Green	Dry	English / Metric
Bending strength	6,110 // 429	9,590 // 674	psi // kgf/cm^2
Density		31 // 496	lbs/ft^3 // kg/m^3
Hardness		320 // 145	lbs // kg
Maximum crushing strength	3,920 // 275	6,340 // 445	psi // kgf/cm^2
Shearing strength		997 // 70	psi // kgf/cm^2
Stiffness	1,020 // 71	1,200 // 84	1,000 psi // 1,000 kgf/cm^2
Weight	30 // 480	24 // 384	lbs/ft^3 // kg/m^3
Radial shrinkage		3	%
Tangential shrinkage		6	%

THUJA OCCIDENTALIS

Description

White cedar (*Thuja occidentalis*) is a particularly slow-growing monoecious tree commonly found in coniferous swamps. It has a columnar crown and a trunk that often divides into secondary stems of similar girth. Mature specimens are about 12 to 15 m (40-50 ft) in height with a trunk diameter of 30 to 60 cm (12 to 24 in). The bark is reddish brown, fissured and peels in narrow strips. Leaves are small and scale-like, closely overlapping. Female flowers are green and solitary with 4 to 6 scales, while male flowers are brown-green and globular. The fruit occurs in the form of cinnamon brown cones enclosing double-winged seeds. The heartwood is light brown, while the narrow sapwood is whitish. Knots in the wood of the white cedar are numerous.

Beschreibung

Der Abendländische Lebensbaum (*Thuja occidentalis*) ist ein sehr langsam wachsender, einhäusiger Baum, den man zumeist in sumpfigen Nadelwäldern findet. Er hat eine säulenartige Krone und einen Stamm, der sich häufig spaltet um einen zweiten Stamm gleichen Umfangs auszubilden. Ältere Exemplare erreichen eine Höhe von etwa 12 bis 15 m, wobei der Stammumfang dann zwischen 30 und 60 cm beträgt. Die Rinde ist in einem rötlichen Braun gehalten, rissig und schält sich in dünnen Streifen ab. Die Blätter sind klein und überlappen sich schuppenartig. Die weiblichen Blüten sind grün und wachsen einzeln mit 4-6 Schuppen, während die männlichen Blüten grünbraun und kugelförmig sind. Die Frucht tritt in Form von zimtbraunen Zapfen auf, die zweiflügelige Samen enthalten. Das Kernholz ist hellbraun, während das Splintholz eine weißliche Farbgebung besitzt. Das Holz des Abendländischen Lebensbaums besitzt zahlreiche Astlöcher.

Scientific / Botanical name
Thuja occidentalis

Trade / Common name
White cedar

Family name
Cupressaceae

Regions / Countries of distribution
Northeast of the United States and southeast of Canada

Global threat status
LC

Common uses

The resistance of white cedar to decay makes it a suitable wood for products in contact with water such as boats, fences, and house siding.

Allgemeine Verwendung

Aufgrund seiner Fäuleresistenz ist das Holz des Thujas besonders für solche Produkte geeignet, die mit Wasser in Kontakt kommen, wie beispielsweise Boote, Zäune oder Hausverkleidungen.

Machinability

Gluing	
Nailing	
Painting	
Screwing	
Staining	
Steam bending	

Physical properties

Numerical data	Green	Dry	English / Metric
Bending strength	4,200 // 295	6,500 // 456	psi // kg/cm²
Hardness		320 // 145	lbs // kg
Impact strength	15 // 38	12 // 30	in // cm
Maximum crushing strength	1,990 // 139	3,960 // 278	psi // kg/cm²
Shearing strength		850 // 59	psi // kg/cm²
Stiffness	640 // 44	800 // 56	1,000 psi // 1,000 kg/cm²
Weight	28 // 448	22 // 352	lbs/ft³ // kg/m³
Radial shrinkage		2	%
Tangential shrinkage		5	%

ABIES LASIOCARPA

Description

Its common name is the alpine fir. It is a medium-sized tree with a height of 15 to 30 m (49-98 ft) and a diameter of 30 to 80 cm (12-31 in). The heartwood and sapwood transition from light yellow to brown tones. The early wood can have lavender or pinkish tones; the knots are yellowish. Its texture is slightly rough to the touch and its grain is nearly homogenous. The wood is malleable, but, in its natural state, not very resistant to environmental factors and especially sensitive to fungi. Marks and scratches are common on the surface of this softwood.

Beschreibung

Auch Felsengebirgstanne oder Felsen-Tanne genannt. Es handelt sich um einen Baum mittlerer Größe, der zwischen 15 und 30 m Höhe und zwischen 30 und 80 cm Stammdurchmesser erreicht. Kern- und Splintholz reichen farblich von hellen Gelbtönen bis hin ins Bräunliche. Das Frühholz kann Farben im Bereich Lavendel oder Rose haben, während die Äste eher ins Gelbliche gehen. Die Textur fühlt sich leicht rau an, die Adern sind fast ebenmäßig. Es handelt sich um ein formbares Holz, das gleichwohl, im natürlichen Zustand, Umweltfaktoren gegenüber sehr sensibel ist. Besonders Pilzbefall gegenüber ist es sehr anfällig. Spuren und Kratzer sind auf Oberflächen, gefertigt aus diesem weichen Holz, keine Besonderheit.

Scientific / Botanical name
Abies lasiocarpa

Trade / Common name
Subalpine fir

Family name
Pinaceae

Regions / Countries of distribution
Western North America

Global threat status
LC

Common uses

It is used as a Christmas tree or as an ornamental tree during the rest of the year. It is commonly used for paper and lightweight construction —for example, as plywood.

Allgemeine Verwendung

Wird als Weihnachtsbaum verwendet und auch zu anderen Anlässen gerne als Zierbaum eingesetzt. Sein Holz ist ein häufig eingesetztes Material in der Papierindustrie, im Leichtbau und in der Herstellung von Sperrholz.

Machinability

Boring	
Gluing	
Mortising	
Moulding	
Nailing	
Planing	
Turning	
Screwing	

Physical properties

Numerical data	Green	Dry	English / Metric
Bending strength	5,050 // 355	8,400 // 590	psi // kgf/cm^2
Hardness		350 // 158	lbs // kg
Maximum crushing strength	2,400 // 168	5,070 // 356	psi // kgf/cm^2
Shearing strength		1,025 // 72	psi // kgf/cm^2
Stiffness	1,155 // 81	1,385 // 97	1,000 psi // 1,000 kgf/cm^2
Radial shrinkage		3	%
Tangential shrinkage		7	%
Volumetric shrinkage		9	%

THUJA PLICATA

Description

The Pacific red cedar or giant red cedar (*Thuja plicata*) is an evergreen coniferous growing about 50 to 70 m (164-230 ft) high. Its often buttressed trunk reaches diameters ranging between 2 and 4 m (6.5-13 ft). The bark is red to gray-brown, fibrous with shallow ridges, and the crown is usually conical, but irregular with pendulous branches. Leaves are scale-like, decurrent and opposite. Two types of cones are borne at the end of branchlets: the seed cones and the pollen cones. The wood of the giant red cedar is reddish brown with occasional darker red or brown streaks. The texture is fairly coarse and the grain is straight with a few knots.

Beschreibung

Der Riesen-Lebensbaum oder Riesen-Thuja (*Thuja plicata*) ist ein immergrüner Nadelbaum, der zwischen 50 und 70 m groß wird. Die oft breit auslaufende Basis seines Stammes kann Durchmesser zwischen 2 und 4 m erreichen. Die Rinde ist graubraun, faserig mit leichten Furchen. Die Krone ist in der Regel kegelförmig, manchmal mit hängenden Zweigen. Die Blätter sind schuppenartig, herablaufend und gegenüberliegend. Am Ende kleiner Zweige kommen zwei Arten von Zapfen vor: Samenzapfen und Pollenzapfen. Das Holz des Riesen-Thujas ist rotbraun, gelegentlich mit dunkelroten oder braunen Streifen durchzogen. Es hat eine recht raue Struktur und eine gerade Maserung mit einigen Astlöchern.

Scientific / Botanical name
Thuja plicata

Trade / Common name
Pacific red cedar

Family name
Cupressaceae

Regions / Countries of distribution
Western North America

Global threat status
LC

Common uses

Thuja plicata is not strong, but it is soft and very resistant to decay. These characteristics make the wood very suitable as construction material for exterior use, such as shingles and fence posts.

Allgemeine Verwendung

Thuja plicata ist nicht stark, sondern weich und sehr fäuleresistent. Diese charakteristischen Eigenschaften machen das Holz zum idealen Baumaterial für den Außenbereich, wie z.B. in Form von Schindeln oder Zaunpfosten.

Machinability

Boring	
Gluing	
Mortising	
Moulding	
Nailing	
Planing	
Polishing	
Splitting	
Staining	
Steam bending	
Turning	

Physical properties

Numerical data	Green	Dry	English / Metric
Bending strength	5,250 // 369	7,700 // 541	psi // kg/cm²
Density		23 // 368	lbs/ft³ // kg/m³
Hardness		350 // 158	lbs // kg
Impact strength	16 // 40	17 // 43	in // cm
Maximum crushing strength	2,770 // 194	4,770 // 335	psi // kg/cm²
Shearing strength		990 // 69	psi // kg/cm²
Stiffness	995 // 70	1,155 // 81	1,000 psi // 1,000 kg/cm²
Weight	28 // 448	23 // 368	lbs/ft³ // kg/m³
Radial shrinkage		2	%
Tangential shrinkage		5	%

CUPRESSUS LUSITANICA

Description

Cupressus lusitanica is an evergreen conifer tree that grows up to 40 m (130 ft) with a trunk diameter of 70 cm (27 in). The cedar of Goa develops a dense conical crown of distinct bluish green foliage. The trunk is short and branches spread out terminating in pendulous branchlets. The bark is thick and rough, reddish brown with longitudinal fissures. This conifer produces ellipsoid cones that are the same color as the foliage when young and that turn reddish brown when mature.

Sapwood is pale usually sharply demarcated from the heartwood. Heartwood is yellowish to pale brown, sometimes variegated. The grain is straight and the texture is fine and uniform.

Beschreibung

Cupressus lusitanica ist eine immergrüne Konifere, die bei einem Stammdurchmesser von 70 cm eine Höhe von bis zu 40 m erreicht. Die Mexikanische Zypresse entwickelt eine dichte kegelförmige Krone mit ausgeprägtem blaugrünen Blattwerk. Aus dem kurzen Stamm ragen Zweige, die sich vergabeln und in kleinen, herabhängenden Zweigen enden. Die dicke, grobe Rinde ist rotbraun mit Längsrissen. Die Zapfen dieser Konifere sind elipsenähnlich geformt und haben bei jungen Bäumen die selbe Farbgebung wie das Blattwerk. Die Zapfen älterer Exemplare werden rotbraun.

Das Splintholz ist blass und in der Regel scharf vom Kernholz abgegrenzt. Das Kernholz ist gelblich bis hellbraun und manchmal mehrfarbig. Die Maserung ist gerade und die Textur fein und gleichmäßig.

Scientific / Botanical name
Cupressus lusitanica

Trade / Common name
Mexican white cedar

Family name
Cunoniaceae

Regions / Countries of distribution
Mexico and Central America

Global threat status
LC

Common uses

The cedar of Goa is used for firewood. Its timber saws cleanly and is used for construction and in the manufacture of furniture.

Allgemeine Verwendung

Die Mexikanische Zypresse wird als Feuerholz verwendet. Das aus ihr gefertigte Schnittholz kann sauber gesägt und für die Baukonstruktion oder zur Herstellung von Mobiliar verwendet werden.

Machinability

Boring	
Gluing	
Mortising	
Moulding	
Nailing	
Painting	
Planing	
Polishing	
Screwing	
Staining	
Steam bending	
Turning	
Veneering qualities	

Physical properties

Numerical data	Green	Dry	English / Metric
Bending strength	6,645 // 467	10,420 // 732	psi // kgf/cm^2
Density		30 // 480	lbs/ft^3 // kg/m^3
Hardness		380 // 172	lbs // kg
Impact strength		30 // 76	in // cm
Maximum crushing strength	3,670 // 258	6,070 // 426	psi // kgf/cm^2
Shearing strength		1,125 // 79	psi // kgf/cm^2
Stiffness	1,160 // 81	1,360 // 95	1,000 psi // 1,000 kgf/cm^2
Weight	28 // 448	24 // 384	lbs/ft^3 // kg/m^3

PICEA ABIES

Description

Picea abies, commonly known as Norway spruce and European spruce, is an evergreen typically growing up to 30 m (98 ft), but in Central Europe heights up to 60 m (197 ft) have been reported. The trunk is generally straight with a thick dark brown bark made of rounded scales that shed easily. Horizontal branches and pendulous branchlets are covered with glossy deep green needles forming a narrow conic crown. Norway spruce produces monoecious flowers and particularly large cones with thin scales that are greenish red when young and maturing brown. Norway spruce wood is creamy white with a tinge of yellow and red, strong, soft and with straight fine grain.

Beschreibung

Picea abies, auch Gemeine Fichte, Gewöhnliche Fichte, Rotfichte oder Rottanne genannt, ist eine immergrüner Baum. Er wächst bis zu 30 m hoch, wobei in Mitteleuropa Höhen von bis zu 60 m dokumentiert sind. Der Stamm ist in der Regel gerade und hat eine dicke, braune Rinde aus rundlichen Schuppen, die leicht abfallen. Die mit glänzend tiefgrünen Nadeln übersäten horizontalen Äste und herabhängenden Zweige bilden eine kegelförmige Krone. Die gemeine Fichte produziert einhäusige Blüten und besonders lange Zapfen, die bei jungen Bäumen rotgrün, bei älteren braun sind. Das Holz der gemeinen Fichte hat ein cremiges Weiß mit einem Hauch Gelb und Rot. Es ist stark, weich und besitzt eine gerade, feine Maserung.

Scientific / Botanical name
Picea abies

Trade / Common name
Norway spruce

Family name
Pinaceae

Regions / Countries of distribution
Europe

Global threat status
LC

Common uses

One of the most economically valuable species in Europe, Norway spruce wood is widely used for construction, furniture and musical instruments.

Allgemeine Verwendung

Als eine der wirtschaftlich wertvollsten Arten Europas wird die Gemeine Fichte weithin in der Baukonstruktion, für die Herstellung von Mobiliar und den Bau von Musikinstrumenten eingesetzt.

Machinability

Boring	
Gluing	
Mortising	
Moulding	
Painting	
Planing	
Polishing	
Staining	

Physical properties

Numerical data	Green	Dry	English / Metric
Bending strength	52,300 // 3,677	9,130 // 641	psi // kgf/cm²
Density		27 // 432	lbs/ft³ // kg/m³
Hardness		380 // 172	lbs // kg
Impact strength	20 // 50	18 // 45	in // cm
Maximum crushing strength	2,815 // 197	5,150 // 362	psi // kgf/cm²
Shearing strength		1,140 // 80	psi // kgf/cm²
Stiffness	1,120 // 78	1,405 // 98	1,000 psi // 1,000 kgf/cm²
Weight	25 // 400	25 // 400	lbs/ft³ // kg/m³
Radial shrinkage		2	%
Tangential shrinkage		7	%

PINUS LAMBERTIANA

Description

The sugar pine, found in Oregon and California, United States, and Baja California, Mexico, is the largest of all of the Pinaceae: it reaches heights of between 30 and 60 m (100-200 ft) and diameters of between 90 cm and 3 m (36-118 in). A member of the white pine subgenus (*Strobus*), it is severely affected by blister rust, a fungus introduced from Europe in 1909. Nevertheless, it is grown and exists in sufficient quantities to be commercialized, normally in a higher class than similar species. Its wood is soft, light and very malleable, and, in general terms, slightly less strong and flexible than that of the yellow pines.

Beschreibung

Die Zucker-Kiefer kommt in Oregon und Kalifornien (USA) und Niederkalifornien (Mexiko) vor. Mit einer Höhe von 30-60 m und einem Stammdurchmesser zwischen 0,9 und 3 m ist sie die größte aller Pinaceae. Als Mitglied der Untergattung der Weymouth-Kiefern (*Strobus*) ist die Zuckerkiefer stark vom Befall durch den Weymouthkiefern-Blasenrost betroffen, einem Pilz, der 1909 aus Europa eingeschleppt wurde. Die Mengen, in denen sie kultiviert wird, reichen jedoch für eine Kommerzialisierung aus; in einer Kategorie, die über der ihrer restlichen Artgenossen liegt. Ihr weiches, leichtes und formbares Holz ist im Allgemeinen weniger stark und elastisch als das der Ocote-Kiefer.

Scientific / Botanical name
Pinus lambertiana

Trade / Common name
Sugar pine

Family name
Pinaceae

Regions / Countries of distribution
Pacific coast of North America

Global threat status
LC

Common uses

Its wood is used for outdoor applications, paneling, roofing, beams, frames, construction materials, carpentry, lightweight construction, flooring, etc.

Allgemeine Verwendung

Ihr Holz eignet sich für die Anwendung im Auenbereich, für die Herstellung von Verkleidungen, Dächern und Dachbalken, Rahmen, Konstruktionsmaterial, Schreinerarbeiten, Leichtbaukonstruktionen, Böden, etc.

Machinability

Boring	
Gluing	
Mortising	
Moulding	
Nailing	
Painting	
Planing	
Polishing	
Sanding	
Staining	
Turning	
Varnishing	

Physical properties

Numerical data	Green	Dry	English / Metric
Bending strength	4,900 // 344	8,200 // 576	psi // kgf/cm²
Hardness		380 // 172	lbs // kg
Impact strength	17 // 43	18 // 45	in // cm
Maximum crushing strength	3,005 // 211	4,460 // 313	psi // kgf/cm²
Shearing strength		1,130 // 79	psi // kgf/cm²
Stiffness	1,030 // 72	1,190 // 83	1,000 psi // 1,000 kgf/cm²
Weight	52 // 832	25 // 400	lbs/ft³ // kg/m³
Radial shrinkage		3	%
Tangential shrinkage		6	%

PINUS STROBUS

Description

The American white pine or Weymouth pine is a species native to North America that reaches between 30 and 40 m (100-130 ft) in height and about 90-120 cm (36-48 in) in diameter. Used as a wood source for boats during the colonial period, the old-growth white pine forests have now been replaced by young specimens. Therefore, even though this is a species that grows rapidly, its supply, practically exhausted in the middle of the 20th century, has been very slow to recover. The wood, virtually identical to that of the sugar pine (*Pinus lambertiana*), is distinguished by its slightly more acidic fragrance and a reduced presence of resin ducts.

Beschreibung

Die Weymouth-Kiefer oder Strobe ist ein nordamerikanischer Endemit, der eine Höhe zwischen 30 und 40 m und einen Stammdurchmesser von 90-120 cm erreicht. Nach ihrer Ausbeutung als Holzquelle für den Schiffbau während der Kolonialzeit, haben sich die alten Baumbestände der Weymouth-Kiefer regeneriert und werden heute durch junge Exemplare ersetzt. Dies erklärt, warum sich ihr Vorkommen, dass Mitte des xx. Jahrhunderts praktisch erschöpft war, trotzdem es sich um eine schnellwachsende Spezies handelt, nur sehr langsam erholt hat. Ihr Holz ist dem der Zucker-Kiefer (*Pinus lambertiana*) quasi gleich und unterscheidet sich nur durch einen etwas säurehaltigeren Duft und dem reduzierten Auftreten von Harzkanälen.

Scientific / Botanical name
Pinus strobus

Trade / Common name
Eastern white pine

Family name
Pinaceae

Regions / Countries of distribution
Eastern North America

Global threat status
LC

Common uses

Carvings, cabinetry, paneling, carpentry, furniture, boats, structural materials, packaging and crates, roofing, musical instruments, moldings, frames, beams and joists, etc.

Allgemeine Verwendung

Schnitzereien, Tischlereiprodukte, Verkleidungen, Schreinerarbeiten, Mobiliar, Schiffbau, Baustoffe, Verpackung und Kisten, Ziegel, Musikinstrumente, Leisten, Rahmen, Dachbalken und -träger, etc.

Machinability

Boring	
Carving	
Gluing	
Mortising	
Moulding	
Nailing	
Planing	
Polishing	
Routing and recessing	
Staining	
Turning	

Physical properties

Numerical data	Green	Dry	English / Metric
Bending strength	4,633 // 325	8,767 // 616	psi // kgf/cm^2
Hardness		380 // 172	lbs // kg
Impact strength	18 // 45	18 // 45	in // cm
Maximum crushing strength	2,333 // 164	4,950 // 348	psi // kgf/cm^2
Shearing strength		900 // 63	psi // kgf/cm^2
Stiffness	970 // 68	1,187 // 83	1,000 psi // 1,000 kgf/cm^2
Weight	41 // 656	26 // 416	lbs/ft^3 // kg/m^3
Radial shrinkage		2	%
Tangential shrinkage		6	%

PICEA ENGELMANNII

Description

The Engelmann spruce is a large conifer with bluish foliage found in eastern North America and northern Mexico, where it exists in isolated populations. Capable of growing to heights of 24-30 m (80-100 ft) and diameters between 45 and 60 cm (18-30 in), it thrives in altitudes that border the alpine tree line and hybridizes with the white spruce (*Picea glauca*) where their habitats overlap. It produces monotone sapwood and heartwood, with a moderately fine texture and straight grain, and with aesthetics and properties that are virtually identical to those of the red spruce (*Picea rubens*) —so much so that they are often sold under the same label.

Beschreibung

Die Engelmann-Fichte ist eine große Konifere mit bläulichem Blattwerk, die im Osten Nordamerikas und im Norden Mexikos beheimatet ist und dort in isolierten Populationen auftritt. Sie kann eine Höhe von 24-30 m und einen Stammdurchmesser von 45-60 cm erreichen. Sie wächst in Höhenlagen, wo sie die alpine Baumgrenze säumt. In Gebieten, in denen sich ihr Habitat mit der Weiß-Fichte (*Picea glauca*) überschneidet, kreuzt sie sich mit dieser. Das Kern- und Splintholz der Engelmann-Fichte unterscheidet sich nicht deutlich. Die Faser ist recht fein und gleichmäßig, die Maserung geradlinig, wobei ihre Ästhetik und Eigenschaften praktisch identisch mit denen der Amerikanischen Rot-Fichte (*Picea rubens*) sind. Daher werden sie oft unter dem selben Etikett verkauft.

Scientific / Botanical name
Picea engelmannii

Trade / Common name
Engelmann spruce

Family name
Pinaceae

Regions / Countries of distribution
Western North America

Global threat status
LC

Common uses

Its resonant properties make it one of the woods most suitable for the manufacture of pianos, violins and xylophones. It is also used in the production of fibers, wood pulp and pellets.

Allgemeine Verwendung

Seine nachhallenden Eigenschaften machen es zu einer der am besten geeigneten Holzarten für die Herstellung von Pianos, Violinen und Xylophonen. Zudem wird es zur Herstellung von Faser, Zellstoff und Pressspan verwendet.

Machinability

Boring	
Gluing	
Mortising	
Moulding	
Nailing	
Planing	
Screwing	
Turning	

Physical properties

Numerical data	Green	Dry	English / Metric
Bending strength	5,200 // 365	9,700 // 681	psi // kgf/cm^2
Density		27 // 432	lbs/ft^3 // kg/m^3
Hardness		390 // 176	lbs // kg
Maximum crushing strength	2,495 // 175	5,300 // 372	psi // kgf/cm^2
Shearing strength		1,200 // 84	psi // kgf/cm^2
Stiffness	1,140 // 80	1,430 // 100	1,000 psi // 1,000 kgf/cm^2
Weight	36 // 576	27 // 432	lbs/ft^3 // kg/m^3
Radial shrinkage		4	%
Tangential shrinkage		8	%

ABIES BALSAMEA

Description

The balsam fir or Christmas fir is a conifer of small to medium size —between 50 and 150 cm in diameter (20-59 in); from 14 to 20 m (46-66 ft) in height, and in exceptional cases, even up to 27 m (89 ft) —whose main features are its conical crown and its fragrant resin. The heartwood and sapwood are easily recognizable. The light tone of the early wood contrasts with the lavender tone of the summerwood. Its texture is fine and uniform, with straight grain and veins that are nearly evenly spaced. It is a softwood, with very limited resistance to weather and parasites, tending to become gray in outdoor conditions.

Beschreibung

Die Balsamtanne ist eine Konifere kleiner bis mittlerer Größe. Sie erreicht einen Stammdurchmesser von 50 bis 150 cm und eine Höhe von 14 bis 20 m, wobei in Ausnahmefällen auch 27 m erreicht werden können. Die Hauptcharaktereigenschaften der Balsam-Tanne sind ihre kegelförmige Krone und ihr duftendes Harz. Kern- und Splintholz der Balsam-Tanne sind sauber voneinander getrennt. Die helle Farbe des Frühholzes hebt sich vom Lavendelton des Spätholzes ab. Seine Textur ist fein und ebenmäßig, mit geradliniger Maserung und regelmäßig angeordneten Adern. Es handelt sich um ein weiches Holz, das dem Klima und Parasiten gegenüber wenig resistent ist. Im Außenbereich wird es gräulich.

Scientific / Botanical name
Abies balsamea

Trade / Common name
Balsam fir

Family name
Pinaceae

Regions / Countries of distribution
Northeastern North America

Global threat status
LC

Common uses

Its resin is used in the production of Canada balsam and turpentine. It serves as a light construction material and is very widely used for packaging and paper.

Allgemeine Verwendung

Das Harz wird zur Herstellung von Kanadabalsam und Terpentin verwendet. Das Holz dient zur Erzeugung von Leichtbaumaterial, Verpackungen und Papierwaren.

Machinability

Boring	
Gluing	
Mortising	
Moulding	
Nailing	
Painting	
Planing	
Staining	
Turning	

Physical properties

Numerical data	Green	Dry	English / Metric
Bending strength	5,400 // 380	8,850 // 622	psi // kgf/cm^2
Density		25 // 400	lbs/ft^3 // kgf/m^3
Hardness		400 // 181	lbs // kg
Impact strength	16 // 41	20 // 51	in // cm
Maximum crushing strength	2,535 // 178	6,125 // 430	psi // kgf/cm^2
Shearing strength		927 // 65	psi // kgf/cm^2
Stiffness	1,190 // 84	1,425 // 100	1,000 psi // 1,000 kgf/cm^2
Weight	45 // 720	25 // 400	lbs/ft^3 // kg/m^3
Radial shrinkage		3	%
Tangential shrinkage		7	%

PINUS MONTICOLA

Description

The western white pine is a species indigenous to North America that generally thrives in the damp soil of mixed forests, at elevations ranging from sea level to 1,000 m (3,500 ft), or up to 3,000 m (9,800 ft) in its more northerly distribution. It grows to heights of about 30 m (100 ft), with diameters of approximately 90 cm (36 in). Soft, stable, moderately heavy and flexible, somewhat brittle and not very resistant, its wood does not serve as a structural component but works very well as a material for carvings or other special uses. Like the sugar pine (*Pinus lambertiana*), it is seriously threatened by blister rust (*Cronartium ribicola*).

Beschreibung

Die Westliche Weymouth-Kiefer, auch Murray-Kiefer genannt, ist in Nordamerika heimisch und wächst in der Regel auf feuchten Böden von Mischwäldern auf Höhe des Meeresspiegels bis hin zu 1 000 m Höhe bzw. 3 000 m Höhe an den nördlichsten Stellen seiner Verbreitung. Die Westliche Weymouth-Kiefer erreicht eine Höhe von ungefähr 30 m und einen Stammdurchmesser von ungefähr 90 cm. Ihr Holz ist weich, beständig, recht schwer und elastisch, etwas brüchig und nicht besonders strapazierfähig. Für die Baukonstruktion ist es ungeeignet, kann aber sehr gut für Schnitzereien und spezifische Verwendungszwecke eingesetzt werden. Wie die Zucker-Kiefer (*Pinus lambertiana*), so ist auch die Westliche Weymouth-Kiefer ernsthaft durch den Weymouthkiefern-Blasenrost (*Cronartium ribicola*) bedroht.

Scientific / Botanical name
Pinus monticola

Trade / Common name
Western white pine

Family name
Pinaceae

Regions / Countries of distribution
Western North America

Global threat status
LC

Common uses
Crates and packaging, interior paneling, matches, flooring, roofing, barrels, construction materials, paneling, moldings, etc.

Allgemeine Verwendung
Kisten und Verpackungen, Polstermöbel, Streichhölzer, Böden, Dächer und Abdeckungen, Fässer, Konstruktionsmaterial, Verkleidungen, Leisten, etc.

Machinability

Boring	
Gluing	
Mortising	
Moulding	
Nailing	
Painting	
Planing	
Polishing	
Staining	
Steam bending	
Turning	
Varnishing	

Physical properties

Numerical data	Green	Dry	English / Metric
Bending strength	4,750 // 333	9,350 // 657	psi // kgf/cm^2
Hardness		420 // 190	lbs // kg
Impact strength	18 // 45	24 // 60	in // cm
Maximum crushing strength	2,465 // 173	5,120 // 359	psi // kgf/cm^2
Shearing strength		1,040 // 73	psi // kgf/cm^2
Stiffness	1,185 // 83	1,455 // 102	1,000 psi // 1,000 kgf/cm^2
Weight	36 // 576	27 // 432	lbs/ft^3 // kg/m^3
Radial shrinkage		4	%
Tangential shrinkage		7	%

PINUS PONDEROSA

Description

The ponderosa pine is a conifer that thrives in mixed and pure forests in the eastern United States and Canada. It has very unusual bark: orange in color and crisscrossed by deep grooves, unique to this species. It grows to heights of between 18 and 39 m (60-130 ft) and diameters of between 80 and 120 cm (30-48 in). With a thick, whitish sapwood, a yellowish or reddish heartwood, a generally straight fiber and uniform texture in which the resin ducts —which are darker— are clearly visible, the wood, which possesses remarkable physical properties —soft, moderately flexible and strong— makes this tree the main timber-producing species in its range of distribution.

Beschreibung

Die Gelb-Kiefer, auch Gold-Kiefer oder Ponderosa-Kiefer genannt, ist eine Konifere, die in Misch- oder reinen Nadelwäldern im Osten der Vereinigten Staaten und Kanada wächst. Sie verfügt über eine spezielle Rinde: Orangefarben, mit tiefen Furchen durchzogen, einmalig unter ihren Artgenossen. Sie wächst bis auf eine Höhe von 18 bis 39 m und hat einen Stammdurchmesser von 80-120 cm. Das dicke Splintholz ist weißlich, das Kernholz gelblich oder leicht rötlich. Seine Faser ist typischerweise geradlinig und die Textur ebenmäßig, sodass die dunkleren Harzkanäle sehr auffallend wirken. Ihr Holz verfügt über beachtliche physische Eigenschaften: Weich, recht elastisch und stark. Diese machen die Gelb-Kiefer zur Hauptspezies für die Nutzholzgewinnung in ihrem Verbreitungsgebiet.

Scientific / Botanical name
Pinus ponderosa

Trade / Common name
Ponderosa pine

Family name
Pinaceae

Regions / Countries of distribution
Western North America

Global threat status
LC

Common uses

Construction materials, carvings, pallets, carpentry, interior paneling, moldings, roofing, furniture, piers, beams, turnery, lightweight construction, etc.

Allgemeine Verwendung

Konstruktionsmaterial, Schnitzereien, Paletten, Schreinerarbeiten, Polstermöbel, Leisten, Ziegel, Mobiliar, Stege, Holzdrehteile, Leichtbaukonstruktionen, etc.

Machinability

- Boring
- Gluing
- Mortising
- Moulding
- Nailing
- Painting
- Planing
- Screwing
- Turning
- Varnishing

Physical properties

Numerical data	Green	Dry	English / Metric
Bending strength	5,400 // 379	9,800 // 689	psi // kgf/cm²
Hardness		460 // 208	lbs // kg
Impact strength	22 // 55	20 // 50	in // cm
Maximum crushing strength	2,645 // 185	5,685 // 399	psi // kgf/cm²
Shearing strength		1,130 // 79	psi // kgf/cm²
Stiffness	1,070 // 75	1,345 // 94	1,000 psi // 1,000 kgf/cm²
Weight	46 // 736	30 // 480	lbs/ft³ // kg/m³
Radial shrinkage		4	%
Tangential shrinkage		6	%

CALOCEDRUS DECURRENS

Description

The California incense cedar is a large conifer —reaching heights of up to 50-60 m (164-197 ft)— native to North America, although it has also spread to Asia and was introduced in Europe in the 19th century, thus making it the most widespread species in its genus. As its name indicates, the brown-orange bark, which turns gray and flaky in old specimens, is used for incense. The heartwood features a clayey tone while the sapwood is the color of cinnamon. The wood is rather durable, though it frequently shows spots of decay caused by fungi. Otherwise, it has excellent properties and, being a softwood, it is ideal for a variety of uses.

Beschreibung

Die Weihrauchzeder, auch Kalifornische Weihrauchzeder oder Kalifornische Flusszeder genannt, ist eine große Konifere, die eine Höhe von 50-60 m erreichen kann. Ursprünglich in Nordamerika beheimatet, wurde sie im XIX. Jahrhundert auch in Europa heimisch gemacht und wird zudem in Asien vertrieben, was sie zu der am weitest verbreiteten Spezies ihrer Gattung macht. Wie ihr Name andeutet, produziert man mit ihrer Rinde Weihrauch. Die Rinde ist braun-orange und wird mit zunehmendem Alter des Baumes grau und beginnt zu schuppen. Das Kernholz der Weihrauchzeder hat die Farbe von Ton, während das Splintholz zimtfarben ist. Es handelt sich um ein recht beständiges Holz, wenn auch häufig durch Pilze verursachte Verrottungsherde auftreten. Abgesehen davon hat das Holz hervorragende Eigenschaften und ist aufgrund seiner weichen Beschaffenheit optimal bearbeitbar.

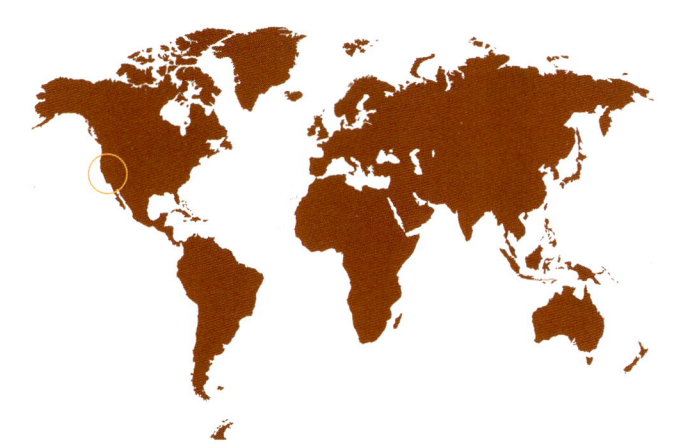

Scientific / Botanical name
Calocedrus decurrens

Trade / Common name
California incense cedar

Family name
Cupressaceae

Regions / Countries of distribution
Western north California

Global threat status
LC

Common uses

Its limited availability is gradually reducing its use as a material. In the past, it was widely used in the manufacture of pencils, and for marquetry, crafts, fencing and shutters.

Allgemeine Verwendung

Seine geringe Verfügbarkeit führt allmählich dazu, dass es immer weniger als Material eingesetzt wird. Früher wurde es häufig für die Herstellung von Bleistiften, Einlegearbeiten, Kunsthandwerk, Zäune und Jalousien verwendet.

Machinability

Boring	
Gluing	
Mortising	
Moulding	
Painting	
Planing	
Polishing	
Staining	
Turning	

Physical properties

Numerical data	Green	Dry	English / Metric
Hardness	390 // 176	470 // 213	lbs // kg
Shearing strength	830 // 58	880 // 61	psi // kgf/cm^2
Radial shrinkage		3.3	%
Tangential shrinkage		5.9	%

ABIES CONCOLOR

Description

Known as the Colorado fir, this medium-to-large-sized conifer ranges from 20 to 50 m (66-164 ft) in height, and has a trunk with a diameter of up to 150 cm (59 in). The heartwood and sapwood are indistinguishable and have a whitish to yellowish color. The wood does not have a specific smell, and is grained with a slightly rough texture, with poor resistance to decomposition and little response to preservatives. The surfaces of the fir tend to be marked with nicks and scratches because the wood is rather soft.

Beschreibung

Die Kolorado-Tanne, auch Grau-Tanne genannt, ist eine mittlere bis große Konifere, deren Höhe zwischen 20 und 50 m schwankt und deren Stamm einen Stammdurchmesser von bis zu 150 cm erreichen kann. Kern- und Splintholz heben sich nicht voneinander ab und haben eine Farbgebung, die zwischen weißlich und gelblich variiert. Das Holz der Kolorado-Tanne ist ohne spezifischen Duft. Es ist gemasert und hat eine leicht raue Struktur. Es ist nur in geringem Maße fäuleresistent und nimmt Schutzanstriche nicht gut auf. Weil es ziemlich weich ist, erleiden Oberflächen aus dem Holz der Kolorado-Tanne verstärkt Kerben und Kratzer.

Scientific / Botanical name
Abies concolor

Trade / Common name
White fir

Family name
Pinaceae

Regions / Countries of distribution
Western North America

Global threat status
LC

Common uses

It is typically used for decorative purposes, though it is also found in paper production and lightweight, temporary and interior construction.

Allgemeine Verwendung

In der Regel dient es zu dekorativen Zwecken, obwohl es auch in der Papierherstellung, beim Leichtbau oder dem Bau temporärer Konstruktionen oder für den Innenausbau Verwendung findet.

Machinability

Boring	
Gluing	
Mortising	
Moulding	
Nailing	
Planing	
Turning	

Physical properties

Numerical data	Green	Dry	English / Metric
Bending strength	5,782 // 406	9,604 // 675	psi // kg/cm²
Hardness		407 // 213	lbs // kg
Impact strength	22 // 55	20 // 50	in // cm
Maximum crushing strength	2,842 // 199	5,684 // 399	psi // kg/cm²
Shearing strength		1,078 // 75	psi // kg/cm²
Stiffness	1,137 // 79	1,470 // 103	1,000 psi // 1,000 kg/cm²
Toughness		162 // 186	in/lbs // cm/kg
Weight	45 // 721	25 // 400	lbs/ft³ // kg/m³
Radial shrinkage	3		%
Tangential shrinkage	7		%
Volumetric shrinkage	11		%

SEQUOIA SEMPERVIRENS

Description

Monotypic genus of plants belonging to the Cupressaceae family. Its only species, the redwood or California redwood, which should not be confused with the giant sequoia (*Sequoiadendron giganteu*) or the dawn redwood (*Metasequoia glyptostroboides*), is a tree of exceptional longevity. It has a lifespan between two and three millennia, on average, and colossal dimensions. With an average height of 60 to 100 m (197-328 ft) —though taller specimens have been recorded— and diameters that range from 3 and 5 m (10-16 ft), it is the largest conifer in the world. It produces a soft wood with a nearly white sapwood and a heartwood the color of mahogany, a straight grain and a texture with wide, clearly visible growth rings.

Beschreibung

Monotypische Pflanzengattung aus der Familie der Cupressaceae. Ihre einzige Spezies, der Küstenmammutbaum oder Küsten Sequoie – nicht zu verwechseln mit dem Riesenmammutbaum (*Sequoiadendron giganteu*) oder dem Urweltmammutbaum (*Metasequoia glyptostroboides*) – ist ein besonders langlebiger Baum, der durchschnittlich zwei bis drei Millennien überdauern kann. Mit einer durchschnittlichen Höhe von 60 bis 100 m – höhere Exemplare sind bekannt – und einem Stammdurchmesser zwischen 3 und 5 m kann von einem Baum mit kolossalen Dimensionen gesprochen werden. Der Küstenmammutbaum ist die größte existente Konifere. Er produziert ein weiches Holz mit fast weißem Splint und einem mahagonifarbenen Kernholz. Die Maserung ist geradlinig und die Textur lässt die Jahresringe deutlich erkennen.

Scientific / Botanical name
Sequoia sempervirens

Trade / Common name
California redwood

Family name
Cupressaceae

Regions / Countries of distribution
Central western coast of the United States

Global threat status
VU

Common uses

Construction materials, cabinetry, moldings, carpentry, posts, roofing, beams, stairs, frames, lightweight construction, boats, furniture, flooring, outdoor features, etc.

Allgemeine Verwendung

Konstruktionsmaterial, Tischlereiprodukte, Leisten, Schreinerarbeiten, Pfosten, Ziegeln, Dachbalken, Treppen, Rahmen, Leichtbaukonstruktionen, Schiffbau, Mobiliar, Böden, Elemente des Außenbereichs, etc.

Machinability

Boring	
Gluing	
Mortising	
Moulding	
Nailing	
Painting	
Planing	
Screwing	
Turning	

Physical properties

Numerical data	Green	Dry	English / Metric
Bending strength	7,500 // 527	10,000 // 703	psi // kg/cm^2
Hardness		480 // 217	lbs // kg
Impact strength	21 // 53	19 // 48	in // cm
Maximum crushing strength	4,200 // 295	6,150 // 432	psi // kg/cm^2
Shearing strength		940 // 60	psi // kg/cm^2
Stiffness	1,180 // 82	1,340 // 94	1,000 psi // 1,000 kg/cm^2
Weight	50 // 800	28 // 448	lbs/ft^3 // kg/m^3
Radial shrinkage		3	%
Tangential shrinkage		4	%

TSUGA CANADENSIS

Description

Tsuga canadensis is a slow-growing coniferous commonly known as eastern hemlock. It reaches heights up to 50 m (164 ft) and can develop a trunk girth of 1.5 m (5 ft). The bark is brown, furrowed and scaly. Its evergreen conical crown is characterized by dark green needles in a spiraled or pectinate (two-ranked) arrangement. The eastern hemlock is a monoecious tree. Male flowers are small, round and yellow and female flowers are green and develop at branch tips. The fruit is in the form of ovoid, light brown cones with scales often projected outward. The wood is soft and coarse-grained with a heartwood that is light red-brown and a sapwood slightly lighter, not clearly demarcated from the heartwood.

Beschreibung

Tsuga canadensis, auch Kanadische Hemlocktanne oder Kanadische Schierlingstanne genannt, ist eine langsamwachsende Konifere. Sie kann bis zu 50 m an Höhe erreichen und einen Stammumfang von 1,5 m entwickeln. Die Rinde ist braun, zerfurcht und schuppig. Die Krone der Kanadischen Hemlocktanne ist immergrün und kegelförmig und zeichnet sich durch seine spiral- oder kammförmig (zweireihig) angeordneten großen, dunklen Nadeln aus. Es handelt sich um einen einhäusigen Baum. Die männlichen Blüten sind klein, rund und gelb, die weiblichen Blüten sind grün und wachsen am Ende der Äste. Die Früchte bestehen aus eiförmigen, hellbraunen Zapfen mit Schuppen, die häufig nach außen ragen. Das Holz ist weich mit rauer Maserung. Die Farbe des Kernholzes ist ein helles Rotbraun. Das Rotbraun des Splintholzes ist noch ein wenig heller und setzt sich nicht klar vom Kernholz ab.

Scientific / Botanical name
Tsuga canadensis

Trade / Common name
Eastern hemlock

Family name
Pinaceae

Regions / Countries of distribution
Eastern North America

Global threat status
LC

Common uses

The primary uses for the timber of *Tsuga canadensis* are limited to pulping, light framing, sheathing, boxes and crates due to its brittleness and numerous knots.

Allgemeine Verwendung

Aufgrund der Brüchigkeit und der vielen Astlöcher beschränkt sich die Hauptverwendung des Schnittholzes der *Tsuga canadensis* auf Aufschluss, leichte Berahmung, Verkleidung, Kisten und Kästen.

Machinability

Gluing	
Nailing	
Planing	
Polishing	
Sanding	
Staining	

Physical properties

Numerical data	Green	Dry	English / Metric
Bending strength	6,600 // 464	9,150 // 643	psi // kg/cm^2
Hardness		500 // 226	lbs // kg
Impact strength	22 // 55	22 // 55	in // cm
Maximum crushing strength	3,175 // 223	5,545 // 389	psi // kg/cm^2
Shearing strength		1,060 // 74	psi // kg/cm^2
Stiffness	1,125 // 79	1,210 // 85	1,000 psi // 1,000 kg/cm^2
Weight	49 // 784	29 // 464	lbs/ft^3 // kg/m^3
Radial shrinkage		3	%
Tangential shrinkage		7	%

PICEA SITCHENSIS

Description

With dimensions ranging from 38 to 53 m (125-175 ft) in height and between 90 and 180 cm (36-72 in) in diameter, the Sitka spruce is the largest species in its genus and the third largest conifer in the world, surpassed only by the California redwood (*Sequoia sempervirens*) and the Douglas fir (*Pseudotsuga menziesii*). Capable of producing large volumes of wood very quickly, this is a major timber-yielding tree —the primary source of wood in Alaska—, and, given its resonant properties, suitable for the production of musical instruments. It is also used as a structural material, as it is an extraordinarily strong wood considering its relative lightness.

Beschreibung

Eine Höhe zwischen 38 und 53 m und ein Stammdurchmesser zwischen 90 und 180 cm machen die Sitka-Fichte zur größten Spezies ihrer Gattung und zur drittgrößten Konifere der Welt. Größer sind nur der Küstenmammutbaum (*Sequoia sempervirerens*) und die Douglasie (*Pseudotsuga menziesii*). Die Sitka-Fichte ist in der Lage, mit hoher Geschwindigkeit große Holzmengen zu produzieren, was sie für die Nutzholzgewinnung zu einer Spezies ersten Ranges und zur Nr. 1-Holzquelle Alaskas macht. Aufgrund seiner nachhallenden Eigenschaften ist es für die Herstellung von Musikinstrumenten geeignet und ist zudem wertvolles Konstruktionsmaterial. Es handelt sich also um ein außerordentlich hartes Holz, das gleichzeitig relativ leicht ist.

Scientific / Botanical name
Picea sitchensis

Trade / Common name
Sitka spruce

Family name
Pinaceae

Regions / Countries of distribution
Western North America

Global threat status
LC

Common uses

It is commonly sold together with other softwoods, making separate identification difficult, except in the case of musical instruments, a niche market in which it has a near monopoly.

Allgemeine Verwendung

In der Regel wird es zusammen mit anderen weißen Holzarten vertrieben, was seine Identifikation erschwert. Dies gilt nicht für Musikinstrumente, einen Markt, auf dem das Holz der Sitka-Fichte beinahe Monopolstellung genießt.

Machinability

Boring	
Gluing	
Mortising	
Moulding	
Nailing	
Painting	
Planing	
Polishing	
Sanding	
Staining	
Steam bending	
Turning	
Varnishing	

Physical properties

Numerical data	Green	Dry	English / Metric
Bending strength	5,300 // 372	10,300 // 724	psi // kgf/cm^2
Hardness		510 // 231	lbs // kg
Impact strength	22 // 55	26 // 66	in // cm
Maximum crushing strength	2,487 // 174	5,555 // 390	psi // kgf/cm^2
Shearing strength		1,150 // 80	psi // kgf/cm^2
Stiffness	1,173 // 82	1,645 // 115	1,000 psi // 1,000 kgf/cm^2
Weight	32 // 512	28 // 448	lbs/ft^3 // kg/m^3
Radial shrinkage		4	%
Tangential shrinkage		8	%

TAXODIUM DISTICHUM

Description

The swamp cypress or bald cypress is a tree of the family Cupressaceae native to the southeastern United States. Considered the symbol of the southern wetlands, it grows in areas that are usually flooded and sunny. It is a large deciduous conifer that reaches heights of about 30-37 m (100-120 ft) and diameters of 90 to 150 cm (36-60 in). It produces a yellowish sapwood and a heartwood with a yellow or reddish color, occasionally blackish. With a straight grain and rough texture, it secretes an oil that makes it practically rot-proof. The wood infected by the *Stereum taxodi* fungus, which features an attractive pattern, is sold under the label *pecky cypress*.

Beschreibung

Die Echte Sumpfzypresse, auch Sumpfeibe genannt, ist ein Baum aus der Familie der Cupressaceae, der aus dem Südosten der Vereinigten Staaten stammt. Als Symbol für die Feuchtgebiete der Südstaaten wächst die Echte Sumpfzypresse in regelmäßig überfluteten, sonnigen Gegenden. Es handelt sich um eine große, sommergrüne Konifere, die eine Höhe von 30-37 m und einen Stammdurchmesser von 90-150 cm erreicht. Sie produziert ein Holz mit gelblichem Splint und rötlich-gelbem, teilweise schwärzlichem Kernholz. Die Maserung ist geradlinig, die Textur rau. Das Öl, das das Holz der Echten Sumpfzypresse absondert, macht es extrem fäuleresistent. Holz, das vom Pilz *Stereum taxodi* befallen ist, erhält ein attraktives Muster und wird unter der Bezeichnung *pecky cypress* vertrieben.

Scientific / Botanical name
Taxodium distichum

Trade / Common name
Bald cypress

Family name
Cupressaceae

Regions / Countries of distribution
Southeastern and Gulf Coastal Plains of the United States

Global threat status
LC

Common uses
Boats, packaging, frames, domestic flooring, furniture, cabinetry, construction materials, planks, carpentry, fuel, etc.

Allgemeine Verwendung
Schiffbau, Verpackungen, Rahmen, Böden, Mobiliar, Tischlereiprodukte, Konstruktionsmaterial, Arbeitsplatten, Schreinerarbeiten, Brennstoff, etc.

Machinability

Boring	
Gluing	
Mortising	
Moulding	
Nailing	
Painting	
Planing	
Polishing	
Turning	

Physical properties

Numerical data	Green	Dry	English / Metric
Bending strength	6,600 // 464	10,500 // 738	psi // kg/cm²
Hardness		510 // 231	lbs // kg
Impact strength	25 // 63	24 // 60	in // cm
Maximum crushing strength	3,580 // 251	6,360 // 447	psi // kg/cm²
Shearing strength		1,000 // 70	psi // kg/cm²
Stiffness	1,180 // 82	1,335 // 93	1,000 psi // 1,000 kg/cm²
Weight	51 // 816	32 // 512	lbs/ft³ // kg/m³
Radial shrinkage		4	%
Tangential shrinkage		6	%

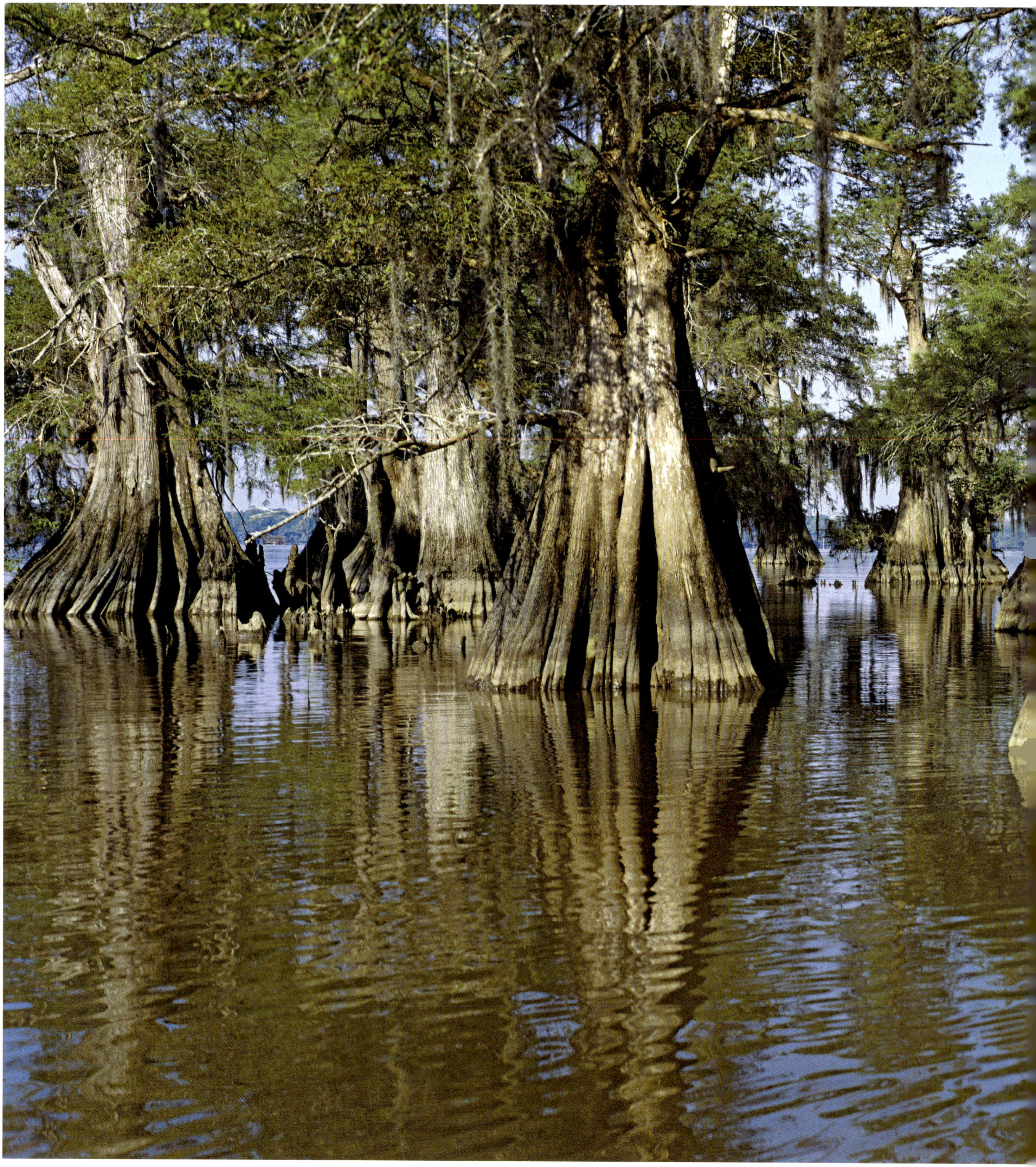

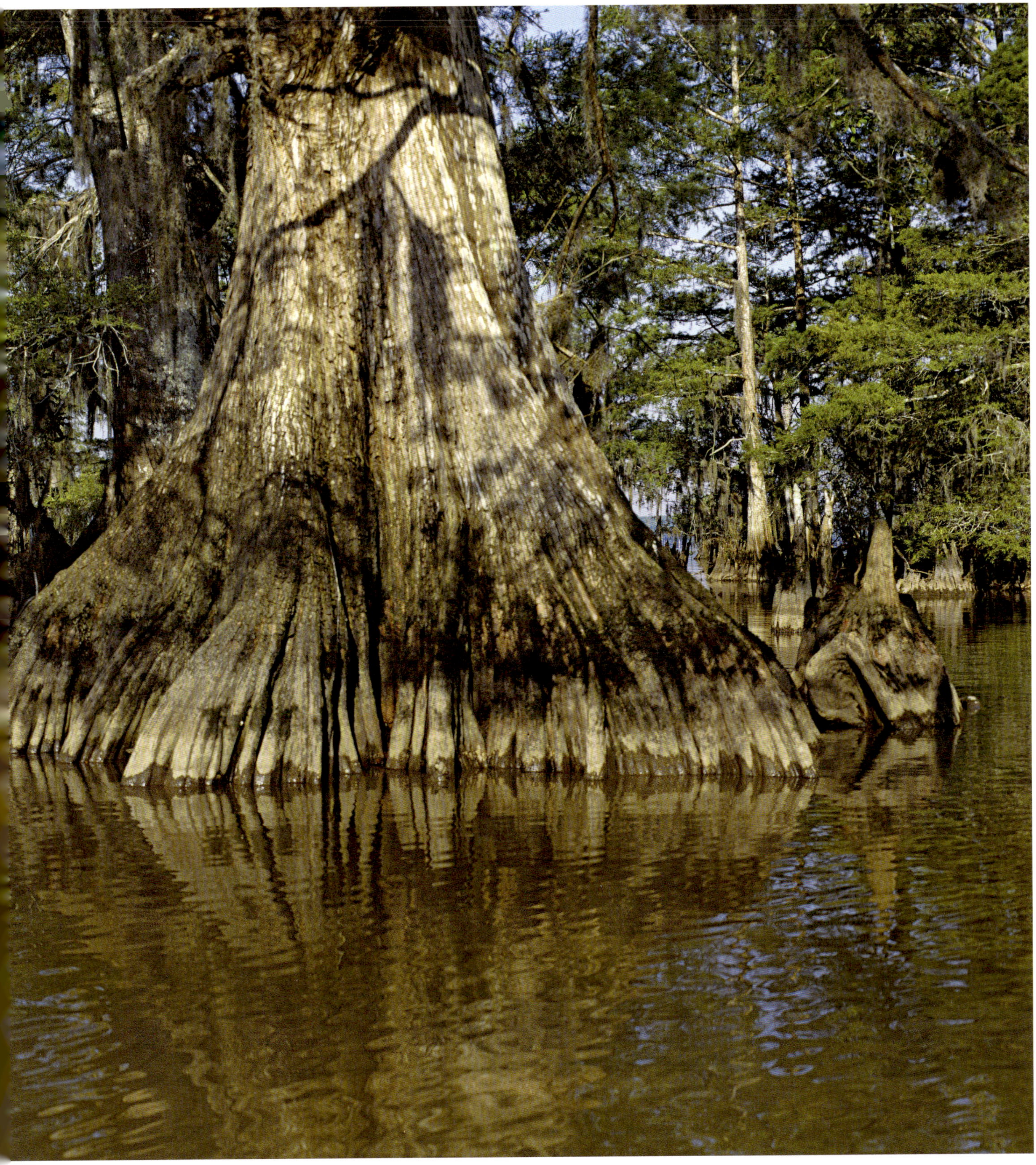

PICEA MARIANA

Description

The black spruce is a small, slow-growing conifer, and these two facts account for its marginal relevance as a timber-yielding tree. Unlike other trees in its genus, which grow in high altitudes, the black spruce thrives in a variety of soils and conditions, thus making it polymorphic: its height can range from 6 m (20 ft) in marshy areas to 18 m (60 ft) in mineral soils. Although the wood is considered stronger than that of the white spruce (*Picea glauca*) or the red spruce (*Picea rubens*), the three are often sold together under the label "eastern spruce or fir". It produces a wood with whitish or cinnamon-colored tones that is somewhat soft, strong and flexible, and rather prone to chipping.

Beschreibung

Die Schwarz-Fichte ist eine kleine, langsamwachsende Konifere. Zwei Eigenschaften, die erklären, warum der Schwarz-Fichte in Bezug auf die Nutzholzgewinnung eine Randbedeutung zufällt. Im Gegensatz zu ihren Artgenossen, die die Höhenlagen bevorzugen, wächst sie auf einer Vielzahl von Böden und unter einer Vielzahl von Konditionen und wird daher als polymorph bezeichnet: Ihre Höhe kann zwischen 6 m in lakustrischen Gebieten und 18 m auf mineralischen Böden schwanken. Obwohl ihrem Holz nachgesagt wird, stärker zu sein als das der Weiß-Fichte (*Picea glauca*) oder das der Rot-Fichte (*Picea rubens*), werden die drei oftmals unter der Bezeichnung „Ost-Fichte oder -Tanne" vermarktet. Das Holz der Schwarz-Fichte ist weißlich oder zimtfarben, ein wenig hart, stark und elastisch und anfällig für Kratzer.

Scientific / Botanical name
Picea mariana

Trade / Common name
Black spruce

Family name
Pinaceae

Regions / Countries of distribution
Northern North America

Global threat status
LC

Common uses

The size of the trunk limits the applicability of its wood to small products like wood pulp, moldings, pellets, crates, pallets, insulation boards and musical instruments.

Allgemeine Verwendung

Die Größe ihres Stammes beschränkt die Verwendungsmöglichkeiten ihres Holzes auf kleine Produkte wie Zellstoff, Leisten, Pressspan, Kisten, Paletten, Dämmplatten oder Musikinstrumente.

Machinability

Boring	
Gluing	
Mortising	
Moulding	
Nailing	
Painting	
Planing	
Screwing	
Turning	

Physical properties

Numerical data	Green	Dry	English / Metric
Bending strength	5,767 // 405	10,550 // 741	psi // kgf/cm^2
Hardness		520 // 235	lbs // kg
Impact strength	24 // 60	24 // 60	in // cm
Maximum crushing strength	2,743 // 192	5,710 // 401	psi // kgf/cm^2
Shearing strength		1,230 // 86	psi // kgf/cm^2
Stiffness	1,243 // 87	1,553 // 109	1,000 psi // 1,000 kgf/cm^2
Weight	36 // 576	30 // 480	lbs/ft^3 // kg/m^3
Radial shrinkage		4	%
Tangential shrinkage		8	%

TSUGA HETEROPHYLLA

Description

Tsuga heterophylla is the western hemlock. It is an evergreen that grows up to 70 m (229.6 ft) with a trunk diameter of 2 m (6.5 ft). Its narrow crown is conical. The bark is gray-brown, becoming deeply furrowed and scaly with age. The notable drooping branches are covered by feathery foliage characterized by two-ranked flat needles. The flowers are monoecious. The numerous cones have flexible scales. Green when young, the cones mature brown with age. The heartwood of the western hemlock is pale brown with an occasional purplish tinge, while the sapwood is almost white to pale brown. The straight-grained wood has an even texture and contains small black knots.

Beschreibung

Tsuga heterophylla, auch Westamerikanische Hemlocktanne oder Westliche Hemlock genannt, ist ein immergrüner Baum. Bei einem Stammdurchmesser von 2 m wächst er bis zu 70 m hoch. Seine schmale Krone ist kegelförmig. Die Rinde ist graubraun und entwickelt im Alter tiefe Furchen und wird schuppig. Die bemerkenswerten herabhängenden Äste sind übersät mit fedrigem Blattwerk, das sich durch seine zweireihigen, flachen Nadeln auszeichnet. Die Blüten sind einhäusig. Die zahlreichen Zapfen haben flexible Schuppen. Die Zapfen sind zunächst grün, werden aber mit zunehmendem Alter braun. Das Kernholz der Westamerikanischen Hemlocktanne ist hellbraun mit gelegentlichen pinken Schattierungen. Das Splintholz hingegen ist fast weiß bis hellbraun. Das Holz besitzt eine gerade Maserung und eine glatte Struktur und enthält kleine schwarze Astlöcher.

Scientific / Botanical name
Tsuga heterophylla

Trade / Common name
Western hemlock

Family name
Pinaceae

Regions / Countries of distribution
West coast of North America

Global threat status
LC

Common uses

Western hemlock is a softwood producing a pale timber with a uniform grain suitable for a wide range of uses including the manufacture of doors, windows and other millwork pieces.

Allgemeine Verwendung

Die Westamerikanische Hemlocktanne ist ein Weichholz aus dem ein blasses Schnittholz mit einheitlicher Maserung entsteht, das in vielen Bereichen Verwendung findet, wie beispielsweise in der Herstellung von Türen, Fenstern und anderen Holzbauteilen.

Machinability

Boring	
Gluing	
Mortising	
Nailing	
Planing	
Polishing	
Resistance to splitting	
Sanding	
Staining	
Steam bending	
Turning	
Veneering qualities	

Physical properties

Numerical data	Green	Dry	English / Metric
Bending strength	6,750 // 474	10,933 // 768	psi // kgf/cm^2
Hardness		540 // 244	lbs // kg
Impact strength	22 // 55	24 // 60	in // cm
Maximum crushing strength	3,460 // 243	6,795 // 477	psi // kgf/cm^2
Shearing strength		1,290 // 90	psi // kgf/cm^2
Stiffness	1,390 // 97	1,660 // 116	1,000 psi // 1,000 kgf/cm^2
Weight	43 // 688	31 // 496	lbs/ft^3 // kg/m^3
Radial shrinkage		5	%
Tangential shrinkage		8	%

CUPRESSUS SEMPERVIRENS

Description

The common cypress or Mediterranean cypress exists in two natural forms: the horizontal variety, with an appearance similar to the cedar, and the pyramidal, which is the most cultivated. It is a species native to the southern Mediterranean and northern Africa, but it has been grown so widely and for so long as an ornamental tree that the majority of its specimens are now found far from its original niche. It produces wood with a light brown color that is knotty and rather light: a material so durable that is considered rot-proof and nearly immune to insects. Though it is not resinous, it emits a scent reminiscent of cedar. It grows rapidly during its first 70 or 80 years and can live more than 1,000 years.

Beschreibung

Die Mittelmeer-Zypresse, auch Säulen-Zypresse, Echte Zypresse, Italienische Zypresse oder Trauer-Zypresse genannt, kommt in zwei natürlichen Formen vor: In der horizontalen Form, ähnlich der Zeder und pyramidenförmig, so wie sie am häufigsten kultiviert wird. Es ist ein Endemit, der im südlichen Mittelmeerraum und in Nordafrika beheimatet ist. Weil er in so großem Ausmaß und seit so langer Zeit als Zierpflanze gezüchtet wird, finden sich die meisten Exemplare der Mittelmeer-Zypresse jedoch fern von ihrem ursprünglichen Lebensraum. Das von ihr produzierte Holz hat eine hellbraune Farbgebung, ist astig und recht leicht: Ein langlebiges Material, das als fäuleresistent bezeichnet wird und so gut wie immun gegenüber Insekten ist. Ohne harzig zu sein verströmt das Holz der Mittelmeer-Zypresse einen Duft, der an Zedern erinnert. Während ihrer ersten 70-80 Lebensjahre ist sie schnellwachsend, wobei sie ein Alter von 1 000 Jahren überschreiten kann.

Scientific / Botanical name
Cupressus sempervirens

Trade / Common name
Mediterranean cypress

Family name
Cupressaceae

Regions / Countries of distribution
Eastern Mediterranean region

Global threat status
NT

Common uses

In the past, its impermeability made the cypress a basic material in shipbuilding. Today, it is used in cabinetry, carpentry, construction, carvings and guitar production.

Allgemeine Verwendung

In der Antike hat ihre Wasserfestigkeit die Zypresse zu einem Basismaterial im Schiffbau gemacht. Heute verwendet man es für Tischlereiprodukte, Schreinerarbeiten, in der Konstruktion, für Bildhauerei und für die Herstellung von Gitarren.

Machinability

Boring	
Gluing	
Mortising	
Nailing	
Planing	
Polishing	
Screwing	
Staining	
Steam bending	

Physical properties

Numerical data	Green	Dry	English / Metric
Density		27 // 432	lbs/ft^3 // kg/m^3
Weight	26 // 416	21 // 336	lbs/ft^3 // kg/m^3

FITZROYA CUPRESSOIDES

Description

Although it is called a larch in its countries of origin (Argentina and Chile), this species is in fact totally unrelated to the common larches, but rather belongs to its own genus, *Fitzroya* (adjectival form of Fitz Roy, who discovered it). Elsewhere, it is known as the Patagonian false cypress. It is a very slow-growing species with extraordinary longevity: the oldest specimens have been around for more than two millennia. Its height varies according to the quality of the soil, ranging from 40 to 70 m (131-230 ft); its diameter is generally between 1.2 and 4.6 m (47-181 in). It produces a soft wood with properties similar to that of the sequoia, but it is not widely commercialized given that it is not intensively cultivated.

Beschreibung

Obwohl sie in den Ländern ihres Ursprungs (Argentinien und Chile) Lärche genannt wird, hat diese Spezies in Wirklichkeit nichts mit der Gemeinen Lärche zu tun, sondern bildet eine Gattung für sich, *Fitzroya* (Adjektivierung von Fritz Roy, ihrem Entdecker). Daher kommt es, dass der Rest sie fälschlicherweise als Patagonische Zypresse kennt. Es handelt sich um eine sehr langsamwachsende Spezies, die extrem langlebig ist: Die ältesten Exemplare zählen mehr als zwei Millennien. Ihre Höhe variiert, je nach Bodenqualität, zwischen 40 und 70 m und ihr Stammdurchmesser schwankt zwischen 1,2 und 4,6 m. Das Holz ist weich und verfügt über Eigenschaften, die denen der Sequoias gleichen. Da sie nicht besonders intensiv gezüchtet werden, hält sich auch ihr Vertrieb in Grenzen.

Scientific / Botanical name
Fitzroya cupressoides

Trade / Common name
Patagonian cypress

Family name
Cupressaceae

Regions / Countries of distribution
Southern Chile and Argentina

Global threat status
EN

Common uses

Musical instruments, carpentry, cigar boxes, pencils, furniture components, wooden roofing, etc.

Allgemeine Verwendung

Musikinstrumente, Schreinerarbeiten, Zigarrenschachteln, Bleistifte, Bauteile für Mobiliar, Holzziegel, etc.

Machinability

Boring	
Gluing	
Mortising	
Moulding	
Planing	
Splitting	
Staining	
Turning	

Physical properties

Numerical data	Green	Dry	English / Metric
Bending strength	6,000 // 421	8,700 // 611	psi // kgf/cm^2
Density		30 // 480	lbs/ft^3 // kg/m^3
Hardness		560 // 254	lbs // kg
Impact strength	36 // 91	45 // 114	in // cm
Maximum crushing strength	2,690 // 189	5,150 // 362	psi // kgf/cm^2
Stiffness	940 // 66	1,160 // 81	1,000 psi // 1000 kgf/cm^2
Radial shrinkage		4	%
Tangential shrinkage		6	%

PINUS RESINOSA

Description

The red pine is a conifer native to North America that takes its common name from its bark, which is thick and ashen at the base and thin and orange closer to the top. It thrives in a wide variety of habitats but maintains great morphological stability, which indicates that it has been on the brink of extinction in its recent evolutionary history. Highly resistant to wind, it exhibits an intolerance to shady areas and a predilection for well-drained soils: in the best conditions, it grows to heights 24 m (80 ft) and can live more than 500 years. Its wood is soft, sturdy and fairly light, though somewhat fragile and susceptible to rotting if not treated with preservatives.

Beschreibung

Die Amerikanische Rot-Kiefer ist eine Konifere mit Ursprung in Nordamerika, die ihren Gattungsnamen ihrer Rinde verdankt. Diese ist an der Basis grob und aschgrau und von der Mitte zur Krone hin fein und orangefarben. Sie wächst in einer großen Vielfalt an Habitaten, bewahrt jedoch eine große morphologische Stabilität, was darauf hinweist, dass sie in ihrer jüngsten Entwicklungsgeschichte kurz vor dem Aussterben stand. Die Amerikanische Rot-Kiefer ist sehr windresistent, meidet düstere Gebiete und hat eine Vorliebe für entwässerte Böden: Unter optimalen Konditionen wächst sie bis zu 24 m hoch und kann ein Alter von 500 Jahren überschreiten. Ihr Holz ist weich, stabil und recht leicht, wenn auch etwas spröde und fäuleanfällig, wenn es keinen Schutzanstrich erhält.

Scientific / Botanical name
Pinus resinosa

Trade / Common name
Red pine

Family name
Pinaceae

Regions / Countries of distribution
North America

Global threat status
LC

Common uses

Crates and packaging, construction materials, pallets, wood pulp, lightweight construction, moldings, carpentry, beams and joists, roofing, sleepers, frames, columns, etc.

Allgemeine Verwendung

Kisten und Verpackungen, Konstruktionsmaterial, Paletten, Zellstoff, Leichtbaukonstruktionen, Leisten, Schreinerarbeiten, Dachbalken und Träger, Ziegel, Eisenbahnschwellen, Rahmen, Pfeiler, etc.

Machinability

Gluing	
Nailing	
Painting	
Screwing	

Physical properties

Numerical data	Green	Dry	English / Metric
Bending strength	5,800 // 407	11,000 // 773	psi // kgf/cm^2
Hardness		560 // 254	lbs // kg
Impact strength	26 // 66	26 // 66	in // cm
Maximum crushing strength	2,730 // 191	6,070 // 426	psi // kgf/cm^2
Shearing strength		1,210 // 85	psi // kgf/cm^2
Stiffness	1,280 // 89	1,630 // 114	1,000 psi // 1,000 kgf/cm^2
Weight	49 // 784	30 // 480	lbs/ft^3 // kg/m^3
Radial shrinkage		4	%
Tangential shrinkage		7	%

PINUS BANKSIANA

Description

The jack pine is a species native to North America with a size generally ranging from 9 to 21 m (30-70 ft) in height and about 30 cm (12 in) in diameter. It is a groundbreaking tree, capable of replenishing forests after fires and logging. It produces a somewhat resinous wood, with a thick, whitish sapwood and an orange, late-forming sapwood —developing after 40 to 50 years of life— with straight grain, clear growth rings and a moderately rough and uniform texture. It is a soft, light and fairly resistant material, though susceptible to rotting, which can be alleviated through the use of preservatives. It is sold together with other Pinaceae —spruces, pines and firs.

Beschreibung

Die Banks-Kiefer ist ein in Nordamerika beheimateter Endemit, dessen Höhe für gewöhnlich zwischen 9 und 21 m schwankt und dessen Stammdurchmesser bei etwa 30 cm liegt. Es handelt sich bei der Banks-Kiefer um eine Pionierpflanze, die die Fähigkeit besitzt, Wälder nach Bränden oder Abholzung wieder aufzuforsten. Sie produziert ein leicht harziges Holz mit grobem, weißlichen Splint und einem Kernholz in Orange, dass sich erst spät entwickelt, wenn der Baum bereits 40 bis 50 Jahre alt ist. Die Holzmaserung ist geradlinig, Wachstumsringe sind klar ausgebildet und die Textur ist recht grob und einheitlich. Es handelt sich um ein weiches Material, das leicht und recht beständig, wenn auch fäuleanfällig ist, was jedoch durch einen Schutzanstrich ausgeglichen werden kann. Man vertreibt das Holz der Banks-Kiefer zusammen mit dem anderer Pinaceae - Fichten, Kiefern, Tannen.

Scientific / Botanical name
Pinus banksiana

Trade / Common name
Jack pine

Family name
Pinaceae

Regions / Countries of distribution
Eastern North America

Global threat status
LC

Common uses

It is common to find its wood in the form of wood pulp and stationery, posts, crates, construction materials, moldings, barrels, roofing, etc.

Allgemeine Verwendung

Für gewöhnlich findet man ihr Holz in Form von Zellstoff, Papierwaren, Pfosten, Kisten, Konstruktionsmaterialien, Leisten, Fässern, Ziegeln, etc.

Machinability

Boring	
Gluing	
Mortising	
Moulding	
Nailing	
Painting	
Planing	
Turning	

Physical properties

Numerical data	Green	Dry	English / Metric
Bending strength	6,000 // 421	9,900 // 696	psi // kgf/cm²
Hardness		570 // 258	lbs // kg
Impact strength	26 // 66	27 // 68	in // cm
Maximum crushing strength	2,950 // 207	5,560 // 390	psi // kgf/cm²
Shearing strength		1,170 // 82	psi // kgf/cm²
Stiffness	1,070 // 75	1,350 // 94	1,000 psi // 1,000 kgf/cm²
Weight	40 // 640	31 // 496	lbs/ft³ // kg/m³
Radial shrinkage		4	%
Tangential shrinkage		7	%

CHAMAECYPARIS NOOTKATENSIS

Description

Although it continues to be sold under the name *Chamaecyparis nootkatensis*, botanists have abandoned this imprecise label in favor of *Callitropsis nootkatensis* and, most recently *Xanthocyparis nootkatensis*. The Nootka (false) cypress —a reference to the native people of Canada who occupied what is now British Columbia— is a very slow-growing tree with extraordinary longevity —the oldest specimens have lived more than 1,500 years. Its size generally ranges from 15 to 30 m (50-100 ft); the trunk measures between 30 and 120 cm (12-48 in). It produces a resinous, soft and resistant wood in external conditions that is greatly valued in the Japanese market.

Beschreibung

Obwohl man im kommerziellen Bereich weiterhin auf die Bezeichnung *Chamaecyparis nootkatensis* zurückgreift, verzichtet man in der Botanik, aufgrund von Ungenauigkeit, darauf und hat sie durch *Callitropsis nootkatensis* bzw. nach neuestem Stand *Xanthocyparis nootkatensis* ersetzt. Die Nootka-Scheinzypresse – in Anspielung auf das kanadische Gebiet ihres Ursprungs, das heute Britisch Kolumbien darstellt – ist ein sehr langsamwachsender Baum von außergewöhnlicher Langlebigkeit. Seine ältesten Exemplare überdauern anderthalb Millennien. Die Höhe der Nootka-Scheinzypresse variiert für gewöhnlich zwischen 15 und 30 m und ihr Stamm hat einen Durchmesser zwischen 30 und 120 cm. Sie produziert ein harziges, weiches und umweltbeständiges Holz, das auf dem japanischen Markt sehr geschätzt wird.

Scientific / Botanical name
Chamaecyparis nootkatensis

Trade / Common name
Yellow cypress

Family name
Cupressaceae

Regions / Countries of distribution
Northwest coast of North America

Global threat status
NE

Common uses

Nautical industry, piers, barrels, greenhouses, cabinetry, furniture, sports equipment, etc.

Allgemeine Verwendung

Schiffsindustrie, Stege, Fässer, Wintergärten, Tischlereiprodukte, Mobiliar, Sportartikel, etc.

Machinability

Boring	
Carving	
Gluing	
Mortising	
Moulding	
Nailing	
Planing	
Polishing	
Resistance to splitting	
Staining	
Steam bending	
Turning	
Veneering qualities	

Physical properties

Numerical data	Green	Dry	English / Metric
Bending strength	6,400 // 449	11,100 // 780	psi // kgf/cm²
Hardness		580 // 263	lbs // kg
Impact strength	27 // 68	29 // 73	in // cm
Maximum crushing strength	3,050 // 214	6,310 // 443	psi // kgf/cm²
Shearing strength		1,130 // 79	psi // kgf/cm²
Stiffness	1,140 // 80	1,420 // 99	1,000 psi // 1,000 kgf/cm²
Weight	36 // 576	31 // 496	lbs/ft³ // kg/m³
Radial shrinkage		3	%
Tangential shrinkage		6	%

LARIX LARICINA

Description

Commonly found in swamps, bogs, and other low-land areas, Tamarack larch is a deciduous coniferous tree with a tapering trunk that rarely grows beyond 20 m (65 ft). The trunk diameter varies between 30 and 60 cm (12 and 24 in). The narrow sapwood is nearly white and the heartwood is light brown. The grain is often twisted and the end grain shows an abrupt transition between early wood and later wood.

Tamarack larch can be easily identified by its small ball-shaped cones.

Beschreibung

Im Allgemeinen in Sümpfen, Mooren und anderen flachlandigen Gegenden zu finden, ist die Ostamerikanische- oder auch Amerikanische Lärche eine sommergrüne Konifere mit einem nach oben hin schmaler werdenden Stamm, die nur selten eine Höhe von 20 m überschreitet. Der Stammdurchmesser variiert zwischen 30 und 60 cm. Das Kernholz ist hellbraun, während das schmale Splintholz eine fast weiße Farbgebung besitzt. Die Maserung ist häufig verdreht und das Hirnholz zeigt einen abrupten Wechsel von Früh- und Spätholz.

Die Tamarack-Lärche, wie sie durch Eindeutschung des englischen Namens auch genannt wird, kann leicht an ihrer kleinen kugelförmigen Zapfen erkannt werden.

Scientific / Botanical name
Larix laricina

Trade / Common name
Tamarack larch

Family name
Pinaceae

Regions / Countries of distribution
Canada

Global threat status
LC

Common uses

The wood of the Tamarack larch is tough but flexible in thin strips. Because of its resistance to rot, the lumber is used for poles. Other common uses include rough lumber and fuel wood.

Allgemeine Verwendung

Das Holz der Tamarack-Lärche ist generell hart, zu dünnen Streifen verarbeitet ist es flexibel. Aufgrund seiner Fäuleresistenz wird das Schnittholz für die Herstellung von Pfosten verwendet. Es wird zudem als Brenn- und Rohholz benützt.

Machinability

Boring	
Mortising	
Moulding	
Planing	
Turning	

Physical properties

Numerical data	Green	Dry	English / Metric
Bending strength	6,850 // 481	10,700 // 752	psi // kg/cm^2
Density		36 // 576	lbs/ft^3 // kg/m^3
Hardness		590 // 267	lbs // kg
Maximum crushing strength	3,270 // 229	6,675 // 469	psi // kg/cm^2
Shearing strength		1,280 // 89	psi // kg/cm^2
Stiffness	1,195 // 84	1,425 // 100	1,000 psi // 1,000 kg/cm^2
Weight	48 // 768	36 // 576	lbs/ft^3 // kg/m^3
Radial shrinkage		3	%
Tangential shrinkage		6	%

PINUS RIGIDA

Description

Native to North America, the pitch pine is a small, fire-resistant conifer —rarely exceeding 18 m (60 ft) in height and 60 cm (12-24 in) in diameter. In the past, it was widely used as a source for pitch and wood for boats, fences and sleepers, which can be accounted for by its resinous quality, which makes it practically rot-proof in external conditions. Today, its importance as a timber-yielding species is marginal because, despite the fact that it is highly adaptable, unlike other domesticated Pinaceae, the pitch pine grows very slowly, develops in multiple trunks that are often stunted and twisted and produces a wood with mediocre physical properties that does not justify its large-scale cultivation.

Beschreibung

Die Pech-Kiefer ist eine kleine, feuerbeständige Konifere, die in Nordamerika beheimatet ist. In seltenen Fällen überschreitet ihre Höhe 18 m und ihr Stammdurchmesser ist meist nicht größer als 60 cm. Einst wurde sie als Pech-Quelle, zum Bau von Schiffen und für die Herstellung von Zäunen und Bahnschwellen verwendet. Erklären kann man dies durch ihre harzige Qualität, die sie im Außenbereich praktisch fäuleresistent macht. Heute hat sie als Spezies zur Nutzholzgewinnung fast gänzlich an Bedeutung verloren. Obwohl sie sehr anpassungsfähig ist, wächst die Pech-Kiefer, im Gegensatz zu anderen, domestizierten Pinaceae, sehr langsam und entwickelt zudem Stammgabelungen, die oft in verkrümmten und verdrehten Stämmen resultieren. Zudem produziert die Pech-Kiefer ein Holz, dessen mittelmäßige, physische Eigenschaften eine Massenkultivierung nicht rechtfertigen.

Scientific / Botanical name
Pinus rigida

Trade / Common name
Pitch pine

Family name
Pinaceae

Regions / Countries of distribution
Eastern North America

Global threat status
LC

Common uses
The wood of the pitch pine is still sold today, but mostly for the purposes of construction material, wood pulp, containers (crates and packaging) or fuel.

Allgemeine Verwendung
Das Holz der Pech-Kiefer findet zwar weiterhin Absatz, dies jedoch hauptsächlich als Konstruktionsmaterial, Zellstoff, Behälter (Kisten und Verpackungen) oder Brennstoff.

Machinability

Boring	
Gluing	
Mortising	
Nailing	
Painting	
Planing	
Polishing	
Screwing	
Sanding	
Staining	
Varnishing	

Physical properties

Numerical data	Green	Dry	English / Metric
Bending strength	6,800 // 478	10,800 // 759	psi // kgf/cm²
Maximum crushing strength	2,950 // 207	5,940 // 417	psi // kgf/cm²
Shearing strength		1,360 // 95	psi // kgf/cm²
Stiffness	1,200 // 84	1,430 // 100	1,000 psi // 1,000 kgf/cm²
Radial shrinkage		4	%
Tangential shrinkage		7	%

PINUS SYLVESTRIS

Description

The Scots pine, Valsaín pine or red pine —a reference to the color of its bark, auburn in the upper part of the trunk and brownish at the base— is a conifer indigenous to northern Eurasia that has now been introduced for ornamental purposes in New Zealand and the United States, where it is considered an invasive species. It regularly reaches heights of about 21 m (70 ft), with trunks measuring approximately 60 cm (24 in) in diameter, though it is not uncommon for it to exceed heights of 30 m (100 ft). Its wide area of distribution accounts for the variability of its physical properties —density and resistance— and aesthetic properties —texture, knottiness— though it can be said that it is a soft and light wood that is generally easy to work with.

Beschreibung

Die Wald-Kiefer, auch Gemeine Kiefer, Rotföhre, Weißkiefer oder Forche genannt, hat ihren Namen auch aufgrund ihrer Rinde, die weiter oben am Stamm scharlachrot und an der Basis bräunlich ist. Es handelt sich um eine bodenständige Konifere aus dem Norden Eurasiens die in Neuseeland und den USA als invasive Art gilt und dort heutzutage zu Zierzwecken eingeführt wird. Regelmäßig werden Höhen von 21 m registriert, wobei es nicht ungewöhnlich ist, dass die Wald-Kiefer gelegentlich auch die 30 m-Marke überschreitet. Ihr Stammdurchmesser liegt bei durchschnittlich 60 cm. Ihr großes Verbreitungsgebiet erklärt, wie breit gefächert Ihre physischen Eigenschaften, wie Dichte und Widerstandsfähigkeit, aber auch ihre ästhetischen Eigenschaften, wie Textur und Knotigkeit, sind. Ihr Holz lässt sich als weich, leicht und einfach bearbeitbar einstufen.

Scientific / Botanical name
Pinus sylvestris

Trade / Common name
Scots pine

Family name
Pinaceae

Regions / Countries of distribution
Europe and Asia

Global threat status
LC

Common uses

Its malleability and lightness have led it to be used for transformative purposes such as veneers, turnery, furniture, carpentry, cabinetry or lightweight construction.

Allgemeine Verwendung

Die Formbarkeit und das geringe Gewicht machen das Holz der Wald-Kiefer zur idealen Wahl, wenn es um transformative Zwecke geht, wie Verkleidungen, Holzdrehteile, Mobiliar, Schreinerarbeiten, Tischlerarbeiten und Leichtbaukonstruktionen.

Machinability

Boring	
Carving	
Gluing	
Mortising	
Moulding	
Planing	
Polishing	
Routing and recessing	
Sanding	
Staining	
Steam bending	
Turning	
Varnishing	

Physical properties

Numerical data	Green	Dry	English / Metric
Bending strength	6,365 // 447	12,255 // 861	psi // kgf/cm²
Hardness		625 // 283	lbs // kg
Impact strength	27 // 68	28 // 71	in // cm
Maximum crushing strength	3,050 // 214	6,595 // 463	psi // kgf/cm²
Shearing strength		1,619 // 113	psi // kgf/cm²
Stiffness	1,130 // 79	1,550 // 108	1,000 psi // 1,000 kgf/cm²
Weight	39 // 624	32 // 512	lbs/ft³ // kg/m³

CHAMAECYPARIS LAWSONIANA

Description

Although it is referred to as the Oregon cedar or the Port Oxford cedar in the horticultural trade, this conifer is actually a cypress, commonly known as the Lawson's cypress. It reaches heights of between 21 and 61 m (70-200 ft) and diameters of 80 to 120 cm (30-48 in). Newly cut, its wood gives off an acrid, gingery smell. Its texture is rough to the touch, with closed pores, straight, streaky grain and uniform veining. The sapwood and heartwood are indistinguishable and exhibit colors that range from a pale yellow to pink or brown tones. Over-exploitation in recent decades has made its wood a very scare commodity, sold mainly in Japan.

Beschreibung

Obwohl sie im Gartenbaugewerbe als Oregonzeder bekannt ist, ist diese Konifere tatsächlich eine Zypresse, die allgemein unter der Bezeichnung Lawsons Scheinzypresse bekannt ist. Sie erreicht eine durchschnittliche Höhe zwischen 21 und 61 m und einen Stammdurchmesser von 80-120 cm. Frisch geschnitten sondert ihr Holz einen scharfen Geruch nach Ingwer ab. Ihre Textur fasst sich rau an, die Poren sind geschlossen, die Maserung geradlinig und gleichmäßig mit Adern durchzogen. Splint- und Kernholz lassen sich nicht unterscheiden und reichen farblich von hellem Gelb bis hin zu Rosa- und Brauntönen. Die Übernutzung der vergangenen Jahrzehnte hat das Holz zu einem seltenen Gut gemacht, dessen Handel sich hauptsächlich auf Japan konzentriert.

Scientific / Botanical name
Chamaecyparis lawsoniana

Trade / Common name
Lawson's cypress

Family name
Cupressaceae

Regions / Countries of distribution
Northwest United States

Global threat status
VU

Common uses

Coffins, beams, construction materials, cabinetry, piers, shipbuilding, posts, frames, furniture, etc.

Allgemeine Verwendung

Särge, Dachbalken, Konstruktionsmaterial, Tischlerarbeiten, Stege, Schiffbau, Pfosten, Rahmen, Mobiliar, etc.

Machinability

Boring	
Gluing	
Mortising	
Moulding	
Nailing	
Planing	
Polishing	
Screwing	
Staining	
Turning	

Physical properties

Numerical data	Green	Dry	English / Metric
Bending strength	5,650 // 397	12,700 // 892	psi // kgf/cm^2
Density		30 // 480	lbs/ft^3 // kg/m^3
Hardness		630 // 285	lbs // kg
Maximum crushing strength	2,565 // 180	6,250 // 439	psi // kgf/cm^2
Shearing strength		1,370 // 96	psi // kgf/cm^2
Stiffness	955 // 67	1,700 // 119	1,000 psi // 1,000 kgf/cm^2
Weight	36 // 576	29 // 464	lbs/ft^3 // kg/m^3
Radial shrinkage		5	%
Tangential shrinkage		7	%

PINUS ECHINATA

Description

This pine, native to the eastern United States, has a great proclivity for polymorphism: with a trunk that is sometimes straight and sometimes twisted and an irregular crown, it reaches heights of between 21 and 30 m (70-100 ft) and trunk diameters of between 45 and 90 cm (18-36 in). It is found in 21 southern states, making it the yellow pine with the widest distribution. Its flaky bark is dark in young specimens but becomes silvery with age. It produces a wood with a straight grain and irregular texture, with a whitish or slightly orange sapwood and a heartwood the color of cinnamon, sometimes very similar to that of the Douglas fir (*Pseudotsuga menziesii*). It is a heavy material that is very flexible and resistant.

Beschreibung

Die Fichten-Kiefer ist im Osten der Vereinigten Staaten beheimatet und tendiert sehr zum Polymorphismus: Der manchmal gerade, manchmal krumme Stamm hat einen Durchmesser von 45 bis 90 cm, die Baumhöhe liegt bei 21 bis 30 m. Die Baumkrone ist unregelmäßig. Sie ist in 21 südlichen Staaten zu finden, was sie zur weitverbreitetsten Gelb-Kiefer macht. Ihre Rinde ist schuppig und an jungen Exemplaren dunkel. Mit zunehmendem Baumalter nimmt die Rinde dann eine silbrige Farbe an. Sie produziert ein Holz mit geradliniger Maserung und unregelmäßiger Textur. Das Splintholz ist weißlich oder leicht orange und das Kernholz ist zimtfarben und teilweise dem der Douglasie (*Pseudotsuga menziesii*) sehr ähnlich. Es handelt sich um ein schweres, sehr elastisches und widerstandsfähiges Material.

Scientific / Botanical name
Pinus echinata

Trade / Common name
Shortleaf pine

Family name
Pinaceae

Regions / Countries of distribution
Eastern United States

Global threat status
LC

Common uses

This is a very commercially important timber-producing species in the construction and carpentry sectors, used for veneers, sleepers, wood pulp, pellets, etc.

Allgemeine Verwendung

Es handelt sich um eine Spezies, die im Bereich der Nutzholzgewinnung von großer kommerzieller Bedeutung ist, was den Bausektor und die Holzverarbeitung betrifft. Zudem findet man ihr Holz in Verkleidungen, Bahnschwellen, Zellstoff, Pressspan, etc.

Machinability

| Boring |
| Gluing |
| Mortising |
| Moulding |
| Nailing |
| Painting |
| Planing |
| Polishing |
| Sanding |
| Screwing |
| Staining |
| Turning |
| Varnishing |

Physical properties

Numerical data	Green	Dry	English / Metric
Bending strength	7,400 // 520	13,100 // 921	psi // kgf/cm²
Hardness		690 // 312	lbs // kg
Impact strength	30 // 76	33 // 83	in // cm
Maximum crushing strength	3,530 // 248	7,270 // 511	psi // kgf/cm²
Shearing strength		1,390 // 97	psi // kgf/cm²
Stiffness	1,390 // 97	1,750 // 123	1,000 psi // 1,000 kgf/cm²
Weight	52 // 832	36 // 576	lbs/ft³ // kg/m³
Radial shrinkage		5	%
Tangential shrinkage		8	%

PINUS TAEDA

Description

The loblolly pine, yellow pine or Rosemary pine is a large conifer —24-30 m (80-100 ft) in height— that is resinous and fragrant and represents the main timber-yielding species of the southern United States. The progressive control of forest fires has allowed the species to spread to areas further south than its natural distribution zone, even taking the place of the longleaf pine (*Pinus palustris*) and the slash pine (*Pinus elliottii*). The physical and aesthetic properties of its wood are comparable to those of the Douglas fir (*Pseudotsuga*), though the taeda pine surpasses it in terms of density and stability.

Beschreibung

Die Weihrauch-Kiefer, auch Amerikanische Terpentin-Kiefer genannt, ist eine große Konifere, von 24-30 m Höhe. Sie ist harzig und wohlriechend und im Süden der Vereinigten Staaten Hauptspezies für die Nutzholzgewinnung. Die allmähliche Kontrolle der Waldbrände hat es der Spezies erlaubt, sich über ihr natürliches Verbreitungsgebiet hinaus, bis in die südlicheren Gebiete auszubreiten. So weit, dass sie sogar beginnt, die Sumpf-Kiefer (*Pinus palustris*) und die Elliott-Kiefer (*Pinus elliottii*) zu ersetzen. Die physischen und ästhetischen Eigenschaften des Holzes sind vergleichbar mit denen der Douglasie (*Pseudotsuga*), wenngleich die Weihrauch-Kiefer diese in Bezug auf Dichte und Stabilität noch übertrifft.

Scientific / Botanical name
Pinus taeda

Trade / Common name
Loblolly pine

Family name
Pinaceae

Regions / Countries of distribution
Southeastern United States

Global threat status
LC

Common uses

The loblolly pine is an important source of firewood and wood pulp, though it is occasionally also found on the market in the form of balustrades, veneers, panels and tool handles.

Allgemeine Verwendung

Die Weihrauch-Kiefer ist eine wichtige Quelle für Brennholz und Zellstoff, gleichwohl man sie auf dem Markt auch in Form von Balustraden, Verkleidungen, Paneelen oder Werkzeuggriffen finden kann.

Machinability

Boring	
Carving	
Gluing	
Mortising	
Moulding	
Nailing	
Planing	
Polishing	
Routing and recessing	
Sanding	
Turning	

Physical properties

Numerical data	Green	Dry	English / Metric
Bending strength	7,300 // 513	12,800 // 899	psi // kgf/cm²
Hardness		690 // 312	lbs // kg
Impact strength	30 // 76	30 // 76	in // cm
Maximum crushing strength	3,510 // 246	7,130 // 501	psi // kgf/cm²
Shearing strength		1,390 // 97	psi // kgf/cm²
Stiffness	1,400 // 98	1,790 // 125	1,000 psi // 1,000 kgf/cm²
Weight	53 // 848	36 // 576	lbs/ft³ // kg/m³
Radial shrinkage		5	%
Tangential shrinkage		7	%

PSEUDOTSUGA MENZIESII

Description

Native to North America, the Douglas fir or Douglasia is the second largest conifer in the world, surpassed only by the sequoia (*Sequoia sempervirens*): its height varies from 24 to 61 m (80-200 ft) with a diameter between 60 and 150 cm (24-60 in). There are two subspecies, the mountain Douglas fir (*glauca* variety), the main timber-producing variety, and the coast Douglas fir (*menziesii* variety). Its wood is in abundant supply due to its wide distribution and very rapid growth. Available at an affordable price, it is a knotless, strong, light material with tones, grain and texture that vary greatly according to the conditions of growth of the species and its variations.

Beschreibung

Die Douglasie, umgangssprachlich auch Douglastanne, Douglasfichte oder Douglaskiefer genannt, ist in Nordamerika heimisch und die zweitgrößte Konifere der Welt. Ausschließlich der Küstenmammutbaum (*Sequoia sempervirens*) ist größer. Die Höhe der Douglasie schwankt zwischen 24 und 61 m, ihr Stammdurchmesser liegt bei 60 bis 150 cm. Sie wird in zwei Varietäten gegliedert, die Gebirgs-Douglasie (var. *glauca*), Hauptspezies für die Nutzholzgewinnung und die Küsten-Douglasie (var. *menziesii*). Da es sich um eine Spezies mit ausgedehntem Verbreitungsgebiet und sehr schnellem Wachstum handelt, ist die Versorgung mit ihrem Holz reichlich gesichert. Es handelt sich um Material, das frei von Astlöchern, stark, leicht und zu erschwinglichen Preisen zu haben ist. In Farbe, Maserung und Textur variabel, abhängig von den Wachstumsbedingungen, denen die Spezies und ihre Varianten ausgesetzt sind.

Scientific / Botanical name
Pseudotsuga menziesii

Trade / Common name
Douglas fir

Family name
Pinaceae

Regions / Countries of distribution
Western North America

Global threat status
LC

Common uses

Crates and packaging, construction materials, flooring, carpentry, plywood, columns, beams and joists, interior paneling, frames, lightweight construction, etc.

Allgemeine Verwendung

Kisten und Verpackung, Konstruktionsmaterial, Böden, Schreinerarbeiten, Sperrholz, Säulen, Dachbalken und -träger, Polstermöbel, Rahmen, Leichtbaukonstruktionen, etc.

Machinability

- Boring
- Gluing
- Mortising
- Moulding
- Nailing
- Planing
- Polishing
- Resistance to splitting
- Sanding
- Staining
- Steam bending
- Turning
- Veneering qualities

Physical properties

Numerical data	Green	Dry	English / Metric
Bending strength	7,682 // 540	12,850 // 903	psi // kgf/cm^2
Hardness		710 // 322	lbs // kg
Impact strength	25 // 63	36 // 91	in // cm
Maximum crushing strength	3,852 // 270	7,465 // 524	psi // kgf/cm^2
Shearing strength		1,130 // 79	psi // kgf/cm^2
Stiffness	1,523 // 107	2,000 // 140	1,000 psi // 1,000 kgf/cm^2
Weight	39 // 624	35 // 560	lbs/ft^3 // kg/m^3
Radial shrinkage		5	%
Tangential shrinkage		8	%

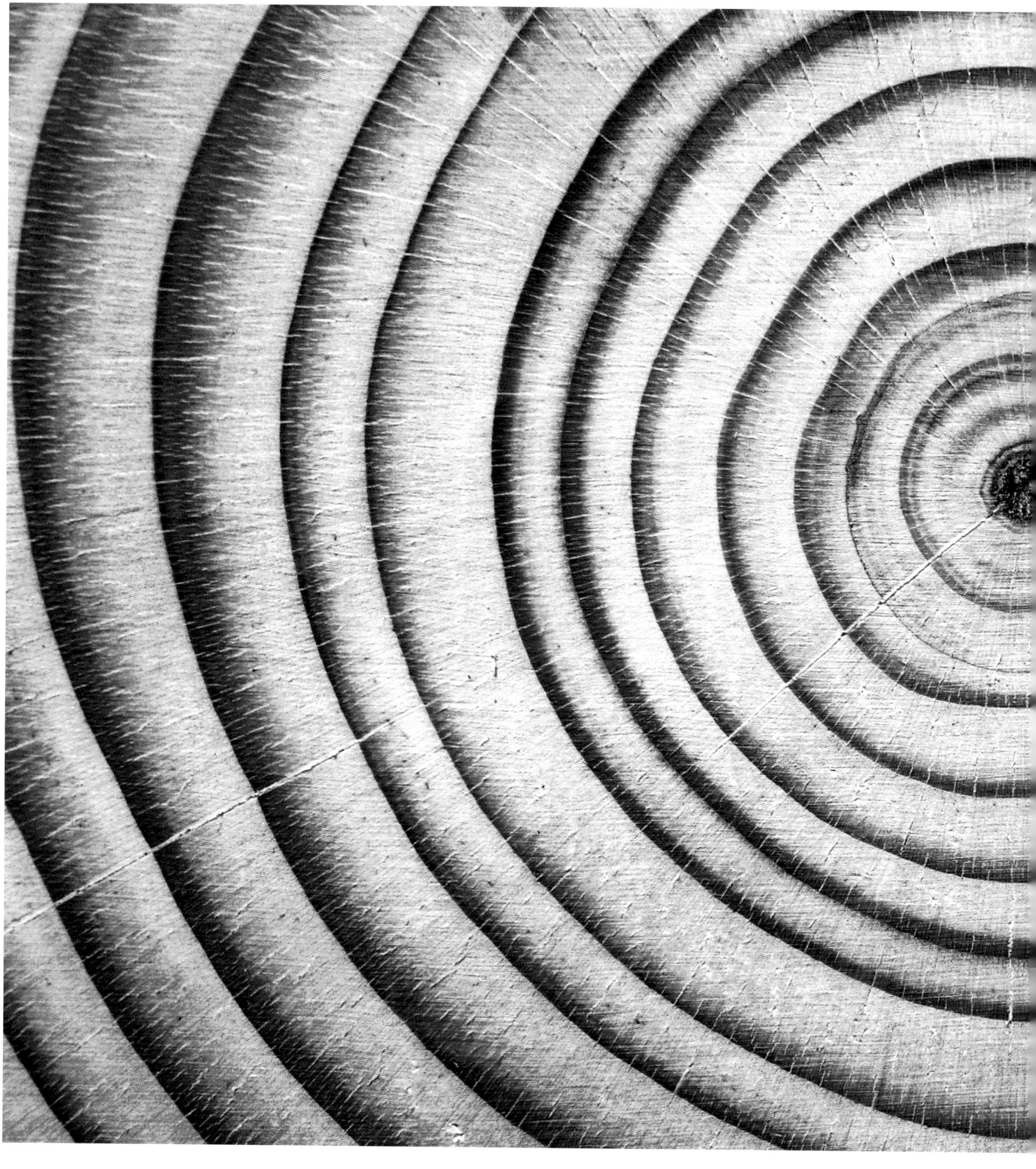

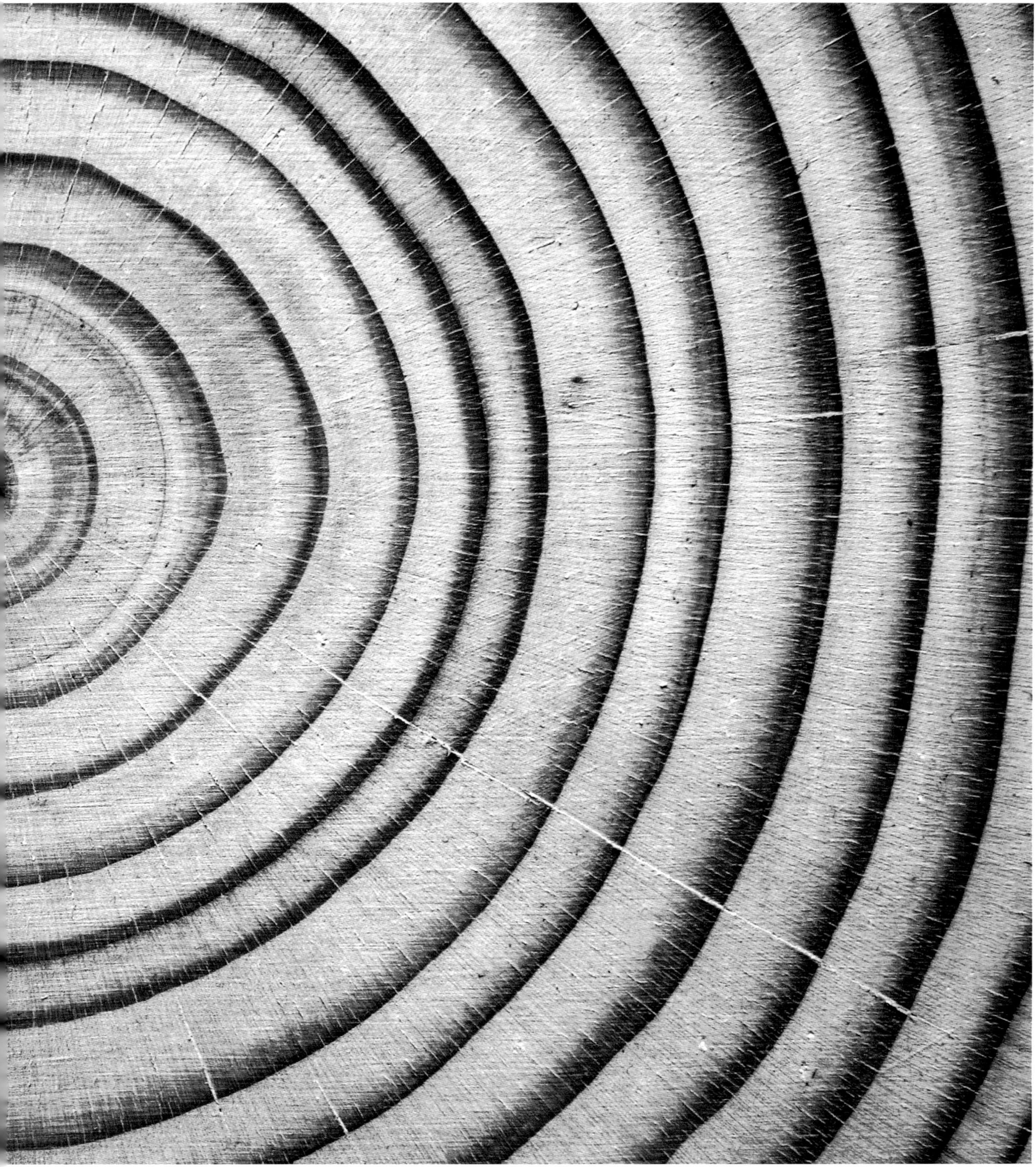

PODOCARPUS GRACILIOR

Description

Fern pine is a small evergreen tree reaching 12 m (40 ft) that has a soft billowy appearance with weeping branch tips, hence its other common name: weeping podocarpus. The trunk grows slowly to be 2 m (6.5 ft) in diameter or larger.

The bark has fine scales, red-brown when young, aging to a light gray. Leaves look more like the leaves of a fern than those of a pine. They are alternate or subopposite, lanceolate, grayish green. Flowers are yellow, not particularly showy. The fruit is round, fleshy and red.

Beschreibung

Das Schlanke Afrogelbholz ist ein kleiner, immergrüner Baum, der bis zu 12 m hoch wird. Mit seinen hängenden Ästen hat er ein sanftes Erscheinungsbild, das ihm im englischen Sprachgebrauch auch den Namen weeping podocarpus, weinender Podocarpus, verleiht. Der Stamm erreicht einen Durchmesser von 2 m oder mehr.

Die Rinde hat feine Schuppen. Bei jungen Exemplaren ist sie rotbraun, bei älteren hellgrau. Die Blätter ähneln eher denen eines Farns als denen einer Pinie. Sie sind wechselständig oder gegenständig, lanzettlich, graugrün. Die Blüten sind gelb und nicht sehr auffallend. Die Frucht ist rund, fleischig und rot.

Scientific / Botanical name
Podocarpus gracilior

Trade / Common name
Fern pine

Family name
Podocarpaceae

Regions / Countries of distribution
Ethiopia, Kenya, Tanzania, and Uganda

Global threat status
NE

Common uses

The timber of the *Podocarpus gracilior* is extensively used in Eastern Africa for construction, paneling, flooring and furniture. In horticulture, it is grown as tree or hedge.

Allgemeine Verwendung

Das Schnittholz des *Podocarpus gracilior* wird in Ostafrika häufig im Bereich der Baukonstruktion, für Verschalungen, als Bodenbelag oder für die Herstellung von Mobiliar eingesetzt. Im Gartenbau wird das Schlanke Afrogelbholz als Baum oder Hecke angepflanzt.

Machinability

Boring	
Carving	
Gluing	
Mortising	
Moulding	
Nailing	
Planing	
Routing and recessing	
Sanding	
Screwing	
Steam bending	
Turning	
Veneering qualities	

Physical properties

Numerical data	Green	Dry	English / Metric
Bending strength	6,335 // 445	9,950 // 699	psi // kg/cm^2
Density		30 // 480	lbs/ft^3 // kg/m^3
Hardness		715 // 324	lbs // kg
Impact strength		24 // 60	in // cm
Maximum crushing strength	3,765 // 264	6,195 // 435	psi // kg/cm^2
Shearing strength		1,150 // 80	psi // kg/cm^2
Stiffness	1,020 // 71	1,200 // 84	1,000 psi // 1,000 kg/cm^2
Weight	31 // 496	25 // 400	lbs/ft^3 // kg/m^3
Radial shrinkage		2	%
Tangential shrinkage		6	%

PINUS ELLIOTTII

Description

The slash pine is a rapidly growing species with a relatively short lifespan —only a couple of centuries, much less than Pinaceae standards— which lives in humid forests in the southeastern United States, where it grows to heights of between 24 and 27 m (80-90 ft) and diameters of between 60 and 75 cm (24-30 in). Its wood has a very attractive appearance, with a sharp transition from the sapwood —thick, ranging from whitish to orange in color— and the heartwood —which has cinnamon or slightly reddish tones— with straight grain, closed pores and very striking patterns. Resinous, heavy, dense, flexible and resistant, it is a first-rate material in the timber industry.

Beschreibung

Die Elliott-Kiefer ist schnellwachsend und in feuchten Wäldern des Südostens der USA beheimatet. Dort erreicht sie eine Höhe zwischen 24 und 27 m und einen Stammdurchmesser zwischen 60 und 75 cm. Es handelt sich um einen relativ kurzlebigen Baum, der kaum ein paar Jahrhunderte alt wird, was, gemessen am Standard der Pinaceae, sehr wenig ist. Das Holz der Elliott-Kiefer ist von hoher Attraktivität. Der Übergang von Splintholz – grob, zwischen weiß und orange – zu Kernholz – zimtfarben oder leicht rötlich – ist schroff. Die Maserung ist geradlinig, die Poren geschlossen, die Muster auffällig. Harzig, schwer, dicht, elastisch, widerstandsfähig und eine Spezies ersten Ranges für die Nutzholzindustrie.

Scientific / Botanical name
Pinus elliottii

Trade / Common name
Slash pine

Family name
Pinaceae

Regions / Countries of distribution
Southeastern United States

Global threat status
LC

Common uses

This species produces one of the most commonly used woods in naval construction. It is also used as a structural material —bridges and beams, for example— or for carpentry, veneers, wood pulp, etc.

Allgemeine Verwendung

Diese Spezies ist Quelle für eine der Holzarten, die am häufigsten im Schiffbau eingesetzt werden. Zudem findet das Holz der Elliott-Kiefer Anwendung als Konstruktionsmaterial – Brücken, Dachträger – oder Schreinerarbeiten, Verkleidung, Zellstoff, etc.

Machinability

Boring	
Gluing	
Mortising	
Moulding	
Nailing	
Painting	
Planing	
Polishing	
Sanding	
Screwing	
Staining	
Turning	
Varnishing	

Physical properties

Numerical data	Green	Dry	English / Metric
Bending strength	8,700 // 611	16,300 // 1,146	psi // kgf/cm^2
Maximum crushing strength	3,820 // 268	8,140 // 572	psi // kgf/cm^2
Shearing strength		1,680 // 118	psi // kgf/cm^2
Stiffness	1,530 // 107	1,980 // 139	1,000 psi // 1,000 kgf/cm^2
Weight	58 // 929	43 // 688	lbs/ft^3 // kg/m^3
Radial shrinkage		5	%
Tangential shrinkage		8	%

ARAUCARIA ANGUSTIFOLIA

Description

The Paraná pine, Brazilian pine or cury´ (its original name in the Guaraní language) is a species at risk of extinction, whose habitat today has been practically reduced to southern Brazil. It reaches heights of between 24 and 37 m (80-120 ft) and diameters of up to 1.5 m (60 in). Its initial shortage, coupled with the costs of distribution by boat, has made this a very expensive wood in the United States and especially in Europe. Its wood, like that of the American white pine, is uniform, with straight grain and barely visible growth rings. The heartwood features reddish stripes with an ocher or chestnut background; the sapwood presents a range of yellows with grayish or orange tones.

Beschreibung

Die Brasilianische Araukarie, auch Brasilkiefer oder Cury´ (ursprünglicher Name auf Guarani) genannt, ist eine vom Aussterben bedrohte Spezies, deren Habitat sich heute praktisch auf den Süden Brasiliens reduziert. Sie wächst bis auf eine Höhe von 24 bis 37 m und hat einen Durchmesser von bis zu 1,5 m. Sein ursprünglicher Mangel, gepaart mit dem Schiffstransport, haben das Holz in den Vereinigten Staaten und speziell in Europa sehr teuer werden lassen. Ähnlich dem der Weymouth-Kiefer ist das Holz der Brasilianischen Araukarie gleichmäßig, mit geradliniger Maserung und kaum sichtbaren Wachstumsringen. Das Kernholz hat rötliche Streifen auf ocker- oder kastanienfarbenem Grund; das Splintholz hat verschiedene Gelbtöne, durchsiebt mit Grau- und Orangetönen.

Scientific / Botanical name
Araucaria angustifolia

Trade / Common name
Brazilian pine

Family name
Araucariaceae

Regions / Countries of distribution
Southern Brazil

Global threat status
CR

Common uses
Beams, packaging, furniture, construction materials, stationery, carpentry, etc.

Allgemeine Verwendung
Dachbalken, Pakete und Verpackungen, Mobiliar, Konstruktionsmaterial, Papierwaren, Schreinerarbeiten, etc.

Machinability

Gluing	
Moulding	
Nailing	
Painting	
Planing	
Polishing	
Staining	
Steam bending	

Physical properties

Numerical data	Green	Dry	English / Metric
Bending strength	7,770 // 546	13,367 // 939	psi // kgf/cm^2
Density		34 // 544	lbs/ft^3 // kg/m^3
Hardness		780 // 353	lbs // kg
Impact strength	23 // 58	24 // 60	in // cm
Maximum crushing strength	3,998 // 281	7,207 // 506	psi // kgf/cm^2
Shearing strength		1,970 // 138	psi // kgf/cm^2
Stiffness	1,378 // 96	1,548 // 108	1,000 psi // 1,000 kgf/cm^2
Radial shrinkage		4	%
Tangential shrinkage		7	%

AGATHIS ROBUSTA

Description

The Queensland kauri (its native state, in Australia) or smooth-barked kauri is a conifer with smooth and flaky bark that grows to a tremendous size —up to 60 m (200 ft) in height with diameters varying between 1.5 and 3 m (60 in-10 ft). It is not an endangered species despite the fact that it was widely harvested in the last two centuries. The sapwood is not distinguishable from the heartwood, which typically features creamy tones ranging from pink to ocher. It has limited natural resistance to decomposition and infestation by insects like shipworms and termites. It is a malleable and robust wood, making it ideal for processing.

Beschreibung

Die Queensland-Kaurifichte (aufgrund ihres australischen Ursprungs) ist eine Konifere mit ebener, schuppiger Rinde. Sie erreicht erstaunliche Höhen von bis zu 60 m und einen Stammdurchmesser zwischen 1,5 und 3 m. Obwohl sie während der letzten zwei Jahrhunderte intensiv abgeholzt wurde, ist die Spezies nicht vom Aussterben bedroht. Das Splintholz hebt sich nicht vom Kernholz ab, das üblicherweise in Cremetönen zwischen rosenfarbig und Ocker gehalten ist. Das Holz der Queensland-Kaurifichte verfügt kaum über eine natürliche Resistenz gegenüber Fäule oder Angriffen durch Insekten wie Muscheln oder Termiten. Es handelt sich um ein formbares und robustes Holz, das aus diesem Grund ideal zu bearbeiten ist.

Scientific / Botanical name
Agathis robusta

Trade / Common name
Queensland kauri

Family name
Araucariaceae

Regions / Countries of distribution
Eastern Queensland in Australia

Global threat status
NE

Common uses
Construction materials, moldings, veneers, plywood, flooring, spools, wood trim, carpentry, etc.

Allgemeine Verwendung
Konstruktionsmaterial, Leisten, Verkleidungen, Sperrholz, Böden, Spulen, Bordüren, Schreinerarbeiten, etc.

Machinability

- Boring
- Carving
- Gluing
- Mortising
- Moulding
- Nailing
- Painting
- Planing
- Polishing
- Staining
- Steam bending
- Turning

Physical properties

Item	Green	Dry	English / Metric
Bending strength	7,487 // 526	13,773 // 968	psi // kgf/cm²
Density		33 // 528	lbs/ft³ // kg/m³
Hardness		785 // 356	lbs // kg
Maximum crushing strength	3,504 // 246	6,833 // 480	psi // kgf/cm²
Shearing strength		2,180 // 153	psi // kgf/cm²
Stiffness	1,442 // 101	1,867 // 131	1,000 psi // 1,000 kgf/cm²
Weight		33 // 528	lbs/ft³ // kg/m³
Radial shrinkage		2	%
Tangential shrinkage		3	%

DACRYDIUM CUPRESSINUM

Description

Once (erroneously) known as the red pine, rimu —a name of Maori origin currently in use — is a true New Zealand conifer. A very slow-growing species, it generally reaches heights of between 24 or 30 m (80-100 ft), occasionally growing as tall as 50 m (164 ft), with trunks measuring up to 200 cm (80 in). Its wood is available in a variety of sizes in New Zealand, though its harvest has been restricted since 1957 as part of a national conservation policy. The highly decorative colored shapes of its dead wood, along with a straight grain, which facilitates lamination, make the rimu an ideal material for veneers.

Beschreibung

Die Rimu-Harzeibe, irrtümlicherweise unter dem Namen Rotkiefer bekannt, heißt in der Sprache der Maori „Rimu". Dieser Name ist auch aktuell in ihrem Ursprungsland, Neuseeland, gebräuchlich. Es handelt sich um eine sehr langsamwachsende Konifere, die für gewöhnlich Höhen zwischen 24 und 30 m erreicht, wobei sie gelegentlich auch die 50 m-Marke knackt. Ihr Stammdurchmesser beträgt bestenfalls 200 cm. Obwohl ihre Abholzung seit 1957 als Teil einer nationalen Erhaltungspolitik beschränkt ist, ist das Holz der Rimu-Harzeibe in Neuseeland in verschiedenen Größen verfügbar. Die pigmentierten, sehr dekorativen Muster des trockenen Holzes in Verbindung mit der geradlinigen Maserung, die das Walzen dessen erleichtert, machen das Holz der Rimu-Harzeibe zu einem optimalen Material für Verkleidungen.

Scientific / Botanical name
Dacrydium cupressinum

Trade / Common name
Rimu

Family name
Podocarpaceae

Regions / Countries of distribution
New Zealand

Global threat status
LC

Common uses

The California pine (*Pinus radiata*) has replaced the rimu in most industries, though the latter continues to be used in high-quality cabinetry and for flooring, plywood and veneers.

Allgemeine Verwendung

Die Monterey-Kiefer (*Pinus radiata*) hat die Rimu-Harzeibe in der Mehrzahl der Industrien ersetzt, obgleich diese ihre Präsenz als Tischlerarbeiten gehobener Qualität und in den Bereichen Bodenbeläge, Sperrholz und Verkleidungen beibehält.

Machinability

Boring	
Gluing	
Nailing	
Painting	
Planing	
Screwing	
Staining	
Turning	

Physical properties

Numerical data	Green	Dry	English / Metric
Bending strength	7,440 // 523	11,100 // 780	psi // kgf/cm^2
Density		37 // 592	lbs/ft^3 // kg/m^3
Hardness		785 // 356	lbs // kg
Maximum crushing strength	3,290 // 231	5,430 // 381	psi // kgf/cm^2
Shearing strength		1,260 // 88	psi // kgf/cm^2
Stiffness	1,220 // 85	1,310 // 92	1,000 psi // 1,000 kgf/cm^2
Weight	60 // 961	37 // 592	lbs/ft^3 // kg/m^3

AGATHIS AUSTRALIS

Description

The kauri is a species of conifer native to New Zealand whose habitat has been reduced today to the northern half of the North Island due to intensive harvesting. Felling of this species is strictly controlled and, as a result, the wood is expensive and relatively difficult to obtain. The diameter of its trunk is exceptional —up to 4.4 m (14.5 ft)— rivaling that of the sequoias. It grows to heights of 50 m (164 ft). It has smooth bark and small, oval-shaped leaves. Its size, hardness and resistance make its wood a popular material for the construction of boats and buildings. The crown and stump of the tree have a highly valued grain that is used for decorating walls and building furniture.

Beschreibung

Der Neuseeländische Kauri-Baum ist eine in Neuseeland beheimatete Konifere, dessen Habitat sich aufgrund intensiver Abholzung heute auf die nördliche Hälfte der Nordinsel reduziert. Seine Ernte wird streng überwacht, was dazu führt, dass sein Holz teuer und relativ schwer zu bekommen ist. Der Durchmesser seines Stammes misst außergewöhnliche 4,4 m Maximum und rivalisiert mit dem der Sequoie. Er wächst aufrecht bis zu einer Höhe von 50 m. Seine Rinde ist eben und seine Blätter klein und oval. Seine Größe, Festigkeit und Widerstandsfähigkeit machten aus dem Holz des Neuseeländischen Kauri-Baums ein beliebtes Material für die Konstruktion von Schiffen und Gebäuden. Die Krone und der Stumpf des Baumes haben eine sehr geschätzte Maserung, die als Wanddekoration und zur Herstellung von Möbeln verwendet wurde.

Scientific / Botanical name
Agathis australis

Trade / Common name
Kauri

Family name
Araucariaceae

Regions / Countries of distribution
New Zealand's North Island

Global threat status
CD

Common uses

As a protected species, its use is restricted. It is currently used extensively for buildings, shipbuilding, bridges, sleepers, moldings, dams, furniture, varnishes, etc.

Allgemeine Verwendung

Aufgrund der Tatsache, dass es sich um eine geschützte Spezies handelt, sind ihre Verwendungsmöglichkeiten begrenzt. Ursprünglich fand das Holz des Neuseeländischen Kauri-Baumes verschwenderisch Verwendung in Gebäuden, Schiffskonstruktionen, Brücken, Eisenbahnschwellen, Formen, Dämmen, Mobiliar, Lacken, etc.

Machinability

Boring	
Gluing	
Mortising	
Moulding	
Painting	
Planing	
Polishing	
Staining	
Turning	

Physical properties

Numerical data	Green	Dry	English / Metric
Hardness		790 // 358	lbs // kg
Maximum crushing strength		6,130 // 430	psi // kg/cm^2
Shearing strength		12,810 // 900	psi // kg/cm^2
Weight		34 // 544	lbs/ft^3 // kg/m^3
Radial shrinkage		4.2	%
Tangential shrinkage		6	%

CEDRUS LIBANI

Description

The Lebanon cedar or Solomon cedar – named after the king who, according to the story, built his temple with wood of this species– is a conifer, native to the mountainous Mediterranean areas, which has been heavily exploited since ancient times. Today, reforestation of the species (Turkey) and conservation of the limited remaining specimens (like the Cedars of God, in Lebanon) are being undertaken. In its natural state, it reaches heights of up to 40 m (130 ft) and diameters of 12 m (40 ft); when cultivated, it rarely exceeds half this size. Its wood, light and smooth-textured, is highly valued as a surface, which, combined with its scarcity, makes it one of the most expensive softwoods on the market.

Beschreibung

Die Libanon-Zeder, auch Salomon-Zeder genannt – König, der der Geschichte nach seinen Tempel aus dem Holz dieser Spezies errichtete – ist eine Konifere. Beheimatet ist sie in den bergigen Gebieten des Mittelmeerraumes, wo sie lange Zeit intensiv ausgebeutet wurde. Heute hat man ihre Wiederaufforstung (Türkei) und den Erhalt ihrer wenigen noch existierenden Reste (wie die Zedern Gottes im Libanon) in Angriff genommen. Im Naturzustand erreicht die Libanon-Zeder eine Höhe von bis zu 40 m und einen Stammdurchmesser von 12 m. In der Züchtung überschreitet sie selten die Hälfte dieser Angaben. Ihr leichtes Holz ist von geschmeidiger Textur und als Verkleidung sehr beliebt. Dies, gepaart mit ihrem knappen Vorkommen, macht sie zu einem der teuersten weichen Holzsorten auf dem Markt.

Scientific / Botanical name
Cedrus libani

Trade / Common name
Lebanon cedar

Family name
Pinaceae

Regions / Countries of distribution
Mediterranean region

Global threat status
LC

Common uses

It is a common ornamental tree in temperate regions. Its wood is an excellent surface and is used in cabinetry. It emits a fragrance similar to incense, which is distilled to make perfumes.

Allgemeine Verwendung

Es handelt sich um eine Zierpflanze, die in gemäßigten Regionen beheimatet ist. Ihr Holz eignet sich hervorragend als Verkleidung und für Tischlerarbeiten. Es sondert einen Duft ab, der dem des Weihrauch ähnlich ist. Durch Destillation werden Parfums hergestellt.

Machinability

Nailing	
Painting	
Planing	
Polishing	
Screwing	
Staining	
Steam bending	
Varnishing	

Physical properties

Numerical data	Green	Dry	English / Metric
Weight		35 // 560	lbs/ft^3 // kg/m^3
Radial shrinkage		4.1	%
Tangential shrinkage		6	%

PINUS RADIATA

Description

Native of the southwestern United States, the insignis pine, Monterrey pine or California pine is the most cultivated softwood species in temperate climates: it can be found in Australia, Brazil and South Africa, but it has been cultivated most extensively and intensively in Chile and New Zealand. It is a rapidly growing tree —mature after 20 years— which, in the right conditions —heat, humidity, and deep, siliceous soil— grows up to 30 m (100 ft) in height and approximately 90 cm (36 in) in diameter. Its wood, which has a straight grain, homogenous texture and very wide sapwood and responds very well to preservatives, is weak and rigid, but there are insignis pine transgenics available to remedy these deficiencies.

Beschreibung

Beheimatet in den Vereinigten Staaten, ist die Monterey-Kiefer die im milden Klima am meisten kultivierte Weichholz-Spezies: Man findet sie in Australien, Brasilien und Südafrika, wobei Chile und Neuseeland die Länder sind, wo die Monterey-Kiefer im größten Umfang und am intensivsten kultiviert wird. Es handelt sich um einen schnellwachsenden Baum, der innerhalb von 20 Jahren heranreift. Unter optimalen Bedingungen, zu denen Hitze, Feuchtigkeit und tiefe, kieselige Böden gehören, wird die Monterey-Kiefer bis zu 30 m hoch und erreicht einen Stammdurchmesser von ungefähr 90 cm. Ihr Holz ist von geradliniger Maserung, homogener Textur und breitem Splint und nimmt Schutzanstriche gut auf. Es ist wenig widerstandsfähig und unelastisch, wenngleich gentechnisch veränderte Varianten der Monterey-Kiefer existieren, die frei von diesen Mängeln sind.

Scientific / Botanical name
Pinus radiata

Trade / Common name
Insignis pine

Family name
Pinaceae

Regions / Countries of distribution
Central coast of California in the United States

Global threat status
CD

Common uses

In Europe, it is used above all as an ornamental tree. Its wood is converted into veneers, crates, lightweight objects, wood pulp, construction materials, etc.

Allgemeine Verwendung

In Europa wird sie vor allem als Zierbaum eingesetzt. Ihr Holz lässt sich zu Verkleidungen, Kisten, Leitbaukonstruktionen, Zellstoff, Konstruktionsmaterialien, etc. verarbeiten.

Machinability

Boring	
Gluing	
Moulding	
Nailing	
Planing	
Polishing	
Staining	
Turning	

Physical properties

Numerical data	Green	Dry	English / Metric
Bending strength	5,700 // 400	11,020 // 774	psi // kgf/cm^2
Hardness		820 // 371	lbs // kg
Impact strength	18 // 45	22 // 55	in // cm
Maximum crushing strength	2,660 // 187	6,038 // 424	psi // kgf/cm^2
Shearing strength		1,700 // 119	psi // kgf/cm^2
Stiffness	1,140 // 80	1,330 // 93	1,000 psi // 1,000 kgf/cm^2
Weight	40 // 640	32 // 512	lbs/ft^3 // kg/m^3
Radial shrinkage		3	%
Tangential shrinkage		7	%

PINUS CONTORTA

Description

The *Pinus contorta* is a species native to western North America, which has spread to places such as Mexico, New Zealand and northern Europe, where it is used in forestry. Its height varies between 6 and 24 m (20-80 ft); its trunk ranges from 30 to 90 cm (12-36 in) in diameter. The common name of the species in English, the lodgepole pine, is a reference to its use in the construction of Native America tepees —cone-shaped tents. The species owes its propagation to fires —the cones need to be exposed to high temperatures to release the seeds. It shares its hardness and flexibility with the rest of the Pinaceae, to the point that it is considered the strongest wood in relation to its weight in the western United States.

Beschreibung

Die Küsten-Kiefer, auch Murray-Kiefer oder Dreh-Kiefer genannt, ist eine im Nordamerikanischen Osten beheimatete Spezies, die heute auch in Mexiko, Neuseeland oder im Norden Europas zu finden ist, wo sie in der Forstwirtschaft verwendet wird. Ihre Höhe schwankt zwischen 6 und 24 m, Ihr Stammdurchmesser beträgt zwischen 30 und 90 cm. Der englische Gattungsname der Spezies *lodgepole*, spielt auf ihre Verwendung beim Tipi-Bau (kegelförmige Zelte) der Indianer an. Ihre Vermehrung legt die Küsten-Kiefer in die Hand der Waldbrände, da ihre Zapfen nur unter großer Hitzeeinwirkung Samen abwerfen. Härte und Elastizität teilt sich die Küsten-Kiefer mit dem Rest der Pinaceae. In einem Punkt hebt sie sich allerdings ab: Ihr Holz wird, in Relation zu seinem Gewicht, als das stärkste des Ostens der Vereinigten Staaten erachtet.

Scientific / Botanical name
Pinus contorta

Trade / Common name
Lodgepole

Family name
Pinaceae

Regions / Countries of distribution
Western North America

Global threat status
LC

Common uses

It is found on the market as a material for construction, panels, decorative veneers, plywood, roofing, wood pulp, sleepers, structural components, etc.

Allgemeine Verwendung

Man findet es auf dem Markt als Konstruktionsmaterial, Paneel, dekorative Verkleidung, Sperrholz, Ziegel, Zellstoff, Bahnschwelle, Bauelement, etc.

Machinability

Boring	
Gluing	
Mortising	
Moulding	
Nailing	
Painting	
Planing	
Polishing	
Screwing	
Staining	
Turning	

Physical properties

Numerical data	Green	Dry	English / Metric
Bending strength	7,100 // 499	12,500 // 878	psi // kgf/cm²
Density		35 // 560	lbs/ft³ // kg/m³
Hardness		850 // 385	lbs // kg
Impact strength	36 // 91	32 // 81	in // cm
Maximum crushing strength	3,040 // 213	6,320 // 444	psi // kgf/cm²
Shearing strength		1,100 // 77	psi // kgf/cm²
Stiffness	1,200 // 84	1,640 // 115	1,000 psi // 1,000 kgf/cm²
Weight	55 // 881	35 // 560	lbs/ft³ // kg/m³
Radial shrinkage		6	%
Tangential shrinkage		10	%

PINUS PALUSTRIS

Description

The longleaf pine, during its herbaceous phases, which consists of the first 5-12 years, seems to be nothing more than a source of green needles; full maturity does not occur until after 100-150 years —by that time, its height is usually around 30 m (100 ft)— and it often lives more than five centuries. It is nearly fireproof and therefore a pioneering tree that, with the passage of time —and, given its longevity, it is not in a hurry— eventually becomes dominant in the forests and savannahs of its range of distribution. Its wood, which is dense, heavy, strong and rather easy to work with, is extensive and intensively cultivated in Georgia, Alabama, Mississippi, Arkansas and Louisiana.

Beschreibung

In den ersten 5-12 Jahren ihres Lebens, in denen sich die Sumpf-Kiefer noch im Unterwuchs befindet, scheint sie nicht mehr, als eine grüne Nadelquelle zu sein. Ihre volle Reife erlangt sie nicht vor einem Alter von 100-150 Jahren, wobei ihre Höhe dann um die 30 m beträgt. Es ist nicht ungewöhnlich, dass sie mehr als fünf Jahrhunderte Lebenszeit überdauert. Gegen Feuer ist sie quasi immun und ist daher ein Pionierbaum, der mit der Zeit – aufgrund seiner Langlebigkeit hat er keine Eile – seine Vorherrschaft über die Wälder und Savannen seines Verbreitungsgebiets einstellt. Sein Holz ist sehr dicht, schwer, stark und recht leicht zu bearbeiten. Seine umfangreiche und intensive Kultivierung findet in Georgia, Alabama, Mississippi, Arkansas und Louisiana statt.

Scientific / Botanical name
Pinus palustris

Trade / Common name
Longleaf pine

Family name
Pinaceae

Regions / Countries of distribution
Southeastern United States

Global threat status
VU

Common uses

Planks, beams, construction materials, sleepers, piers and jetties, veneers and plywood, flooring, wood pulp, structural components, boats, etc.

Allgemeine Verwendung

Arbeitsplatten, Dachträger, Konstruktionsmaterial, Eisenbahnschwellen, Stege und Dämme, Verkleidungen und Sperrholz, Böden, Zellstoff, Bauelemente, Schiffbau.

Machinability

Boring	
Gluing	
Mortising	
Moulding	
Nailing	
Painting	
Planing	
Polishing	
Sanding	
Staining	
Steam bending	
Turning	
Varnishing	

Physical properties

Numerical data	Green	Dry	English / Metric
Bending strength	8,500 // 597	14,500 // 1,019	psi // kgf/cm^2
Hardness		870 // 394	lbs // kg
Impact strength	35 // 88	34 // 86	in // cm
Maximum crushing strength	4,320 // 303	8,470 // 595	psi // kgf/cm^2
Shearing strength		1,510 // 106	psi // kgf/cm^2
Stiffness	1,590 // 111	1,980 // 139	1,000 psi // 1,000 kgf/cm^2
Weight	55 // 881	41 // 656	lbs/ft^3 // kg/m^3
Radial shrinkage		5	%
Tangential shrinkage		8	%

JUNIPERUS VIRGINIANA

Description

The red cedar or juniper is a slow-growing coniferous evergreen that can reach a height of 12 to 18 m (40 to 60 ft) and develop a trunk above buttresses of 30 to 60 cm (12 to 24 in). *Juniperus virginiana* develops two types of leaves: small dark green scale-like and blue-green needle-like leaves. The fruit is berry-like, originally green, turning a dark bluish color when reaching maturity. The sapwood is pale yellow and the heartwood is generally reddish or purple-brown and very fragrant. The grain is straight, often knotty, and has a very fine texture with close pores.

Beschreibung

Der Virginische Wacholder, auch Bleistiftzeder, Virginische Zeder oder Virginische Rotzeder genannt, ist eine langsamwachsende, immergrüne Konifere, die 12 bis 18 m Höhe erreichen kann und deren Stamm, oberhalb seiner breit auslaufenden Basis, 30 bis 60 cm Durchmesser hat. Der Virginische Wacholder hat zwei Arten von Blättern: Kleine, dunkelgrüne, schuppenartige und blaugrüne nadelähnliche Blätter. Die Frucht ist beerenähnlich. Erst grün, später ein dunkler Blauton. Das Splintholz hat ein helles Gelb und das Kernholz ist in der Regel rötlich oder lila-braun und stark duftend. Die Maserung ist gerade, oftmals mit Astlöchern und hat eine sehr feine Struktur mit geschlossenen Poren.

Scientific / Botanical name
Juniperus virginiana

Trade / Common name
Red cedar

Family name
Cupressaceae

Regions / Countries of distribution
Eastern North America

Global threat status
LC

Common uses

Because of its resistance to rotting, the wood of the Red cedar is used for fence posts. It is often marketed as "aromatic cedar" because of its fragrant wood that is commonly used as lining for closets and also repels moths. Juniper oil is distilled from the wood and leaves.

Allgemeine Verwendung

Aufgrund seiner Fäuleresistenz wird das Holz der Virginischen Rotzeder häufig für die Herstellung von Zaunpfosten verwendet. Nicht selten wird sie aufgrund ihres duftenden Holzes als „Aromatische Zeder" vermarktet und für die Auskleidung von Schränken und zur Mottenbekämpfung eingesetzt. Juniper Öl wird aus dem Holz und den Blättern destilliert.

Machinability

Boring	
Mortising	
Moulding	
Planing	
Polishing	
Turning	

Physical properties

Numerical data	Green	Dry	English / Metric
Bending strength	7,000 // 492	8,800 // 618	psi // kgf/cm^2
Density		33 // 528	lbs/ft^3 // kg/m^3
Hardness		900 // 408	lbs // kg
Impact strength	35 // 88	22 // 55	in // cm
Maximum crushing strength	3,570 // 250	6,020 // 423	psi // kgf/cm^2
Shearing strength		1,010 // 71	psi // kgf/cm^2
Stiffness	650 // 45	880 // 61	1,000 psi // 1,000 kgf/cm^2
Weight	37 // 592	33 // 528	lbs/ft^3 // kg/m^3
Radial shrinkage		3	%
Tangential shrinkage		5	%

PINUS CARIBAEA

Description

The Caribbean pine is a species native to the tropical and subtropical conifer forests of Central America and part of the Caribbean (Cuba and the Bahamas, where two island varieties of the species exist), though its distribution now includes areas such as Jamaica, Colombia, South Africa, China and India, where it grows to heights of about 30 m (100 ft), with trunks measuring approximately 75-100 cm (30-40 in) in diameter. It tolerates acidic and low-quality soils, which favors its cultivation for production of wood pulp and softwood where other timber-yielding species are incapable of thriving. Its wood is dense, heavy, sturdy, malleable and flexible.

Beschreibung

Die Karibik-Kiefer ist in den tropischen und subtropischen Nadelwäldern Zentralamerikas und Teilen der Karibik (Kuba und die Bahamas, wo zwei Insel-Varietäten der Spezies existieren) beheimatet. Ihr exotisches Verbreitungsgebiet umfasst heute jedoch ebenfalls Jamaika, Kolumbien, Südafrika, China und Indien, wo sie Höhen bis zu 30 m und einen Stammdurchmesser von 75-100 cm erreicht. Dass sie sogar auf sauren und kargen Böden wächst, wirkt sich überall dort begünstigend auf ihre Kultivierung zur Gewinnung von Zellstoff und Weichholz aus, wo andere Spezies zur Nutzholzgewinnung unfähig sind zu wachsen. Das Holz der Karibik-Kiefer ist dicht, schwer, stabil, formbar und elastisch.

Scientific / Botanical name
Pinus caribaea

Trade / Common name
Caribbean pine

Family name
Pinaceae

Regions / Countries of distribution
Central America, Cuba, the Bahamas, and the Turks and Caicos Islands

Global threat status
LC

Common uses

It is used in cabinetry and production of flooring, bridges, sleepers, construction materials, furniture, beams and joists, wood pulp, etc.

Allgemeine Verwendung

Man verwendet das Holz der Karibik-Kiefer für Tischlerarbeiten, Böden, Brücken, Eisenbahnschwellen, Konstruktionsmaterial, Mobiliar, Dachbalken und -träger, Zellstoff, etc.

Machinability

Boring	
Carving	
Gluing	
Mortising	
Moulding	
Nailing	
Planing	
Routing and recessing	
Sanding	
Turning	

Physical properties

Numerical data	Green	Dry	English / Metric
Bending strength	10,260 // 721	14,725 // 1,035	psi // kgf/cm^2
Hardness		1,120 // 508	lbs // kg
Impact strength	40 // 101	36 // 91	in // cm
Maximum crushing strength	4,850 // 340	8,540 // 600	psi // kgf/cm^2
Shearing strength		2,090 // 146	psi // kgf/cm^2
Stiffness	1,880 // 132	2,240 // 157	1,000 psi // 1,000 kgf/cm^2
Weight	61 // 977	48 // 768	lbs/ft^3 // kg/m^3
Radial shrinkage		6	%
Tangential shrinkage		8	%

TAXUS BACCATA

Description

The English yew or simply yew is an evergreen conifer that can live for centuries. It grows 10 to 20 m (33-65.6 ft) high with a trunk diameter up to 2 m (6.5 ft). It is heavily branched with thin, scaly bark. Leaves are linear, dark green, and spirally-arranged. Unlike most conifer trees, the yew does not bear its seeds in a cone, but rather in a fleshy red aril at the tip of a shoot. The sapwood of the yew is a thin tan band clearly demarcated from the heartwood, which is orange-brown with occasional darker brown or purple streaks. It is considered the hardest of all hardwoods with a fine grain and smooth texture.

Beschreibung

Die Europäische Eibe, auch Gemeine Eibe oder einfach nur Eibe genannt, ist eine immergrüne Konifere, die jahrhundertealt werden kann. Bei einem Stammdurchmesser von bis zu 2 m wächst sie 12 bis 20 m hoch. Die Eibe hat sehr viele Äste und eine dünne, schuppige Rinde. Ihre Blätter haben eine linealische Form, sind dunkelgrün und spiralförmig angeordnet. Ungleich der meisten Koniferen trägt die Eibe ihre Samen nicht in einem Zapfen, sondern in einem fleischigen, roten Arillus an der Spitze eines Triebes. Das Splintholz der Eibe ist ein dünnes bräunliches Band, das sich deutlich vom Kernholz abhebt. Das Kernholz hat eine orange-braune Farbgebung, gelegentlich mit dunkleren braunen oder lila Streifen durchzogen. Eibenholz gilt als härtestes aller Harthölzer und besitzt eine feine Maserung und glatte Struktur.

Scientific / Botanical name
Taxus baccata

Trade / Common name
Yew

Family name
Taxaceae

Regions / Countries of distribution
Western, central and southern Europe, northwest Africa, northern Iran and southwest Asia

Global threat status
LC

Common uses

Currently, the principal use of the yew is as an ornamental plant. In the past, it was used for cogs, axle-trees and pulley pins because of its hardness, and its colorful wood was used to veneer furniture and for various art and religious objects.

Allgemeine Verwendung

Gegenwärtig findet die Eibe hauptsächlich als Zierpflanze Verwendung. Aufgrund seiner Härte verwendete man ihr Holz in der Vergangenheit zur Herstellung von Zahnrädern, Wagenachsen und Teile von Flaschenzügen. Wegen des Farbenreichtums wurde Eibenholz zudem als Möbelfurnier und für verschiedene Kunst- und Kirchenobjekte verwendet.

Machinability

Boring	
Gluing	
Nailing	
Planing	
Polishing	
Staining	
Steam bending	
Turning	

Physical properties

Numerical data	Green	Dry	English / Metric
Density		40 // 640	lbs/ft^3 // kg/m^3
Weight	40 // 640	32 // 512	lbs/ft^3 // kg/m^3
Radial shrinkage		2	%
Tangential shrinkage		5	%

AGATHIS DAMMARA

Description

The *Agathis dammara* is a resinous tree commonly known as the Borneo kauri. Its long list of synonyms —more then twenty scientific names have been assigned to it— can be explained by its former importance in the trade of the resin copal; *dammar*, in fact means "copal". Although it has been intensively harvested, it continues to be commercially exploited today. Its wood is very dense, soft and as flexible as teak. Its texture is fine, uniform and shiny. It grows to heights of 45 m (150 ft) and diameters of between 2 and 4 m (6,5-13 in). While its appearance and properties are virtually identical to those of *Agathis borneensis*, they are completely distinct species.

Beschreibung

Der *Agathis dammara* ist ein harziger Baum, der allgemein als Dammarabaum bekannt ist. Seine enorme Synonymie – man kann ihm mehr als zwanzig wissenschaftliche Namen zuweisen – liegt in seiner ursprünglichen Relevanz für den Handel mir Kopalharz begründet; *dammar* bedeutet tatsächlich „Kopal". Obwohl er Opfer intensiver Abholzungen wurde, wird er heute weiterhin kommerziell ausgebeutet. Sein Holz ist dicht, leicht und so elastisch wie das Teak. Seine Textur ist fein, gleichmäßig und glänzend. Der Dammarabaum wächst bis auf eine Höhe von 45 m und hat einen Durchmesser von 2-4 m. Auch wenn sein Aussehen und seine Eigenschaften ihn identisch mit *Agathis borneensis* erscheinen lassen, handelt es sich um eine komplett unterschiedliche Spezies.

Scientific / Botanical name
Agathis dammara

Trade / Common name
Damar

Family name
Araucariaceae

Regions / Countries of distribution
The Moluccas, and the Philippines

Global threat status
NE

Common uses

Construction of boats, decorative plywood, veneers, carpentry, turnery, balustrades, flooring, construction materials, automotive components, stairs, etc.

Allgemeine Verwendung

Schiffbau, dekoratives Sperrholz, Verkleidungen, Schreinerarbeiten, Holzdrehteile, Balustraden, Böden, Konstruktionsmaterial, Automobilkomponenten, Treppen, etc.

Machinability

Boring	
Gluing	
Mortising	
Moulding	
Planing	
Polishing	
Turning	

Physical properties

Numerical data	Green	Dry	English / Metric
Weight		42 // 672	lbs/ft^3 // kg/m^3
Radial shrinkage		4.1	%
Tangential shrinkage		7.6	%

BURSERA SIMARUBA

Description

Known by numerous native names, the copperwood or gumbo-limbo is a species indigenous to semiarid areas and savannahs of the Americas. It reaches heights of between 15 and 18 m (50-60 ft), occasionally up to 27 m (90 ft), and diameters of between 60 and 90 cm (24-36 in). In Cuba and Florida, it is known as the tourist tree due to the coppery color of its bark, which also peels off. It produces an aromatic resin that is used in varnishes and incense. While its wood is abundant due to the range of its distribution, it has little commercial value as a result of its poor physical properties. It serves as a soil stabilizer or urban tree due to its ability to thrive in all types of soil.

Beschreibung

Der Weißgummibaum oder Amerikanische Balsam ist unter einer Vielzahl einheimischer Namen bekannt und stellt eine eigene Spezies der semiariden Klimazonen und der amerikanischen Savanne dar. Er wächst bis auf eine Höhe von 15 bis 18 m, manchmal sogar bis zu 27 m hoch und hat einen Stammdurchmesser von 60-90 cm. In Kuba und Florida kennt man ihn als Touristenbaum, weil seine Rinde eine kupferige Farbe hat, zudem schuppt sie sich. Er produziert ein aromatisches Harz, das für Lacke oder Weihrauch verwendet wird. Aufgrund seiner schlechten Materialeigenschaften ist das Holz des Weißgummibaumes trotz seines großen Verbreitungsgebietes und damit üppigen Vorhandenseins auf dem Markt nur von geringem kommerziellen Wert. Verwendung findet das Holz bei der Stabilisierung von Böden. Dank seiner Fähigkeit, auf jeder Art von Boden zu wachsen, wird der Weißgummibaum zudem bei der Bepflanzung von Städten eingesetzt.

Scientific / Botanical name
Bursera simaruba

Trade / Common name
Gumbo-limbo

Family name
Burseraceae

Regions / Countries of distribution
Tropical regions of the Americas

Global threat status
NE

Common uses

This species can be found on the American market as low-quality plywood or surfacing. It is also used in carpentry or the production of packaging, matches, posts, stakes, etc.

Allgemeine Verwendung

Auf dem amerikanischen Markt ist das Holz des Weißgummibaumes in Form von Sperrholz oder Verkleidung minderer Qualität zu finden. Zudem verwendet man es für Schreinerarbeiten oder zur Herstellung von Verpackungen, Streichhölzern, Pfosten, Pflöcken, etc.

Machinability

Boring	
Carving	
Gluing	
Mortising	
Moulding	
Nailing	
Planing	
Polishing	
Resistance to splitting	
Sanding	
Staining	
Turning	
Veneering qualities	

Physical properties

Numerical data	Green	Dry	English / Metric
Bending strength	3,218 // 226	4,902 // 344	psi // kgf/cm^2
Density		24 // 384	lbs/ft^3 // kgf/m^3
Hardness		323 // 146	lbs // kg
Maximum crushing strength	1,831 // 128	3,363 // 236	psi // kgf/cm^2
Shearing strength		882 // 62	psi // kgf/cm^2
Stiffness	796 // 55	1,066 // 74	1,000 psi // 1,000 kgf/cm^2
Weight	24 // 384	20 // 320	lbs/ft^3 // kg/m^3
Radial shrinkage		2	%
Tangential shrinkage		4	%

AESCULUS GLABRA

Description

The *Aesculus glabra* or Ohio buckeye is a species native to North America. It reaches heights of 10 to 25 m (32-82 ft) with diameters of about 60 cm (24 in). A species with little commercial value, its wood is often sold mixed with others. Its flowers, light green in color and with stamens longer than the petals, are rich in tannic acid and thus toxic to humans and livestock. Its fruit is a prickly seedpod measuring 4-5 cm (1.6-2 in) similar to a walnut. Its wood is almost white, sometimes crisscrossed by faint gray streaks, with a straight or occasionally wavy grain. It is a low-density wood with notable resistance given its modest weight.

Beschreibung

Der *Aesculus glabra*, auch Ohio-Rosskastanie genannt, ist eine in Nordamerika heimische Spezies. Die Ohio-Rosskastanie erreicht Höhen zwischen 10 und 25 m und einen Stammdurchmesser von ca. 60 cm. Ihr Holz erzielt einen geringen Marktpreis und wird häufig gemischt mit anderen Holzsorten verkauft. Ihre Blüten, hellgrün und mit Staubblättern, die die Blütenblätter an Länge überragen, sind reich an Gerbsäure und daher für Mensch und Tier giftig. Ihre stachelige Kapselfrucht von 4-5 cm ist ähnlich einer Walnuss. Ihr Holz ist fast weiß und nur stellenweise von feinen grauen Adern durchzogen. Die Maserung ist geradlinig oder manchmal wellig. Es handelt sich um ein Holz mit geringer Dichte, das jedoch aufgrund seines geringen Gewichtes eine bemerkenswerte Festigkeit aufweist.

Scientific / Botanical name
Aesculus glabra

Trade / Common name
Ohio buckeye

Family name
Sapindaceae

Regions / Countries of distribution
Midwestern United States

Global threat status
NE

Common uses

Crates and packaging, carvings, wood pulp, turnery, household items and wooden utensils.

Allgemeine Verwendung

Kisten und Verpackungen, Schnitzereien und Skulpturen, Zellstoff, Holzdrehteile, Haushaltsartikel und Holzgerätschaften.

Machinability

Boring	
Carving	
Mortising	
Moulding	
Planing	
Polishing	
Staining	
Turning	

Physical properties

Item	Green	Dry	English / Metric
Density		32 // 512	lbs/ft^3 // kgf/m^3
Hardness		350 // 158	lbs // kg
Maximum crushing strength		4,170 // 293	psi // kgf/cm^2
Weight		25 // 400	lbs/ft^3 // kg/m^3
Radial shrinkage		3.6	%
Tangential shrinkage		8.1	%

ACER NEGUNDO

Description

Known as the silver maple in the United States, the Manitoba maple in Canada and the acezintle in Mexico, the *Acer negundo* distinguishes itself from similar species by its pinnate leaves, measuring 12 to 15 cm in length (5-6 in) with three, five or as many as seven leaflets. It is a dioecious tree, meaning that it has differentiated sexes, which is also unique among maples. It is the smallest —9-18 m (30-59 ft) in height and about 80 cm (31 in) in diameter— and lightest of the American maples. Its short lifespan and limited resistance to wind and frost make this tree unsuitable for use as an ornamental tree. It is an easily cultivated and rapidly growing species that has become an affordable and versatile commercial wood despite its poor properties.

Beschreibung

Bekannt als Silber-Ahorn in den Vereinigten Staaten, als Manitoba-Ahorn in Kanada und als Acezintle in Mexiko, hebt sich der *Acer negundo* durch seine Fiederblätter von seinen Artgenossen ab. Diese haben eine Länge von 12 bis 15 cm und bestehen aus drei, fünf oder bis zu sieben Einzelblättchen. Der Eschen-Ahorn ist zweihäusig getrenntgeschlechtlich und eine weitere Spezifikation unter den Ahornen. Mit einer Höhe von 9-18 m und einem Stammdurchmesser von ca. 80 cm ist er der Kleinste und Leichteste unter den amerikanischen Ahornen. Seine kurze Lebensdauer und seine geringe Widerstandsfähigkeit gegenüber Wind und Eis sind Gründe dafür, dass er sich nicht als Zierpflanze eignet. Es ha ndelt sich um eine Spezies, die sich leicht kultivieren lässt und schnell wächst. Trotz negativer Eigenschaften ist sein Holz erschwinglich und vielseitig verwendbar.

Scientific / Botanical name
Acer negundo

Trade / Common name
Silver maple

Family name
Sapindaceae

Regions / Countries of distribution
North America

Global threat status
NE

Common uses

It is common in urban parks or high-traffic areas due to its ability to absorb dioxides and its rapid growth. It is used for cabinetry, flooring, paneling and veneer.

Allgemeine Verwendung

Aufgrund seines schnellen Wachstums und seiner Fähigkeit, Dioxide zu absorbieren, ist er üblicherweise in städtischen Parkanlagen oder verkehrsreichen Zonen zu finden. Sein Holz verwendet man für Tischlerarbeiten, Böden, Polstermöbel, Furniere und Verkleidungen.

Machinability

Nailing	
Planing	
Polishing	
Staining	
Steam bending	

Physical properties

Numerical data	Green	Dry	English / Metric
Bending strength	5,220 // 367		psi // kgf/cm²
Density		30 // 480	lbs/ft³ // kg/m³
Hardness		720 // 326	lbs // kg
Maximum crushing strength		4,950 // 348	psi // kgf/cm²
Weight		34 // 544	lbs/ft³ // kg/m³
Radial shrinkage		4	%
Tangential shrinkage		7.5	%

AESCULUS HIPPOCASTANUM

Description

The horse-chestnut is often confused with the buckeye because of their similar fruits. In this case, unlike the species of the genus *Castanea*, its fruit is unfit for human consumption (the term *hippocastanum*, literally "horse chestnut", refers to its former use as livestock feed). Its white flowers are grouped in pyramid-shaped clusters. At a mature age, the diameter of the trunk measures about 60 cm (24 in) and its height exceeds 20 m (66 ft). Its texture is smooth and uniform. The sapwood is confused with the heartwood, the color of which varies between white tones in specimens harvested at the beginning of winter and straw-colored tones in those harvested later.

Beschreibung

Aufgrund der Ähnlichkeit ihrer Früchte wird die Gewöhnliche Rosskastanie häufig mit der Falschen Kastanie verwechselt. In diesem Fall sind ihre Früchte im Gegensatz zu den Spezies der Gattung *Castanea* nicht für den menschlichen Verzehr geeignet (der Begriff *hippocanastum*, wörtlich „Pferdekastanie", weist auf ihre frühere Verwendung als Viehfutter hin). Ihre Blüten sind in pyramidenförmigen Büscheln zusammengefasst. Im mittleren Lebensalter beträgt ihr Stammdurchmesser in etwa 60 cm und ihre Größe überschreitet 20 m. Ihre Textur ist geschmeidig und gleichmäßig. Der Splint vermischt sich mit dem Kernholz, dessen Farbe zwischen verschiedenen Weißtönen schwankt, wenn der Baum zu Beginn des Winters geschlagen wurde und verschiedenen Gelbtönen, wenn der Baum später geschlagen wurde.

Scientific / Botanical name
Aesculus hippocastanum

Trade / Common name
Horse-chestnut

Family name
Sapindaceae

Regions / Countries of distribution
Southeast Europe

Global threat status
NE

Common uses

Pharmacology (anti-inflammatories, vasoprotectants), medicinal uses (circulatory disorders), plywood, moldings, wrapping, interior decorations, construction materials, turnery, etc.

Allgemeine Verwendung

Pharmakologie (entzündungshemmende Mittel, Vasoprotektoren), medizinische Verwendung (Kreislaufstörungen), Sperrholz, Leisten, Verpackungen, Innendekor, Konstruktionsmaterial, Holzdrehteile, etc.

Machinability

Boring	
Gluing	
Mortising	
Moulding	
Nailing	
Planing	
Polishing	
Sanding	
Staining	
Steam bending	
Turning	
Veneering qualities	

Physical properties

Item	Green	Dry	English / Metric
Bending strength	5,605 // 394	9,785 // 687	psi // kgf/cm^2
Density		32 // 512	lbs/ft^3 // kg/m^3
Hardness		750 // 340	lbs // kg
Maximum crushing strength	2,429 // 170	5,616 // 394	psi // kgf/cm^2
Shearing strength		1,221 // 85	psi // kgf/cm^2
Stiffness	824 // 57	974 // 68	1,000 psi // 1,000 kgf/cm^2
Weight	41 // 656	32 // 512	lbs/ft^3 // kg/m^3
Radial shrinkage		2	
Tangential shrinkage		3	

SWIETENIA MACROPHYLLA

Description

A member of the Meliaceae family, the big leaf mahogany is a tree native to tropical regions and is one of the main timber-producing species existing today. In fact, it is in such demand that there is no guarantee that, under the current conditions of exploitation, its supply can be sustained at the existing levels. It grows to heights of between 40 and 50 m (131-164 ft) and diameters measuring about 180 cm (72 in). Its wood is notable both for its aesthetic qualities —its lustrous texture and its color of red wine (so special that the mahogany color is named after it)— and for its clear patterns and physical properties —it is a durable, resistant and exceptionally malleable material.

Beschreibung

Mitglied der Familie der Meliaceae. Beim Amerikanischen Mahagoni handelt es sich um einen Baum, der charakteristisch in tropischen Regionen zu finden ist und heutzutage eine der wichtigsten Spezies zur Nutzholzgewinnung darstellt. Tatsächlich ist er derart gefragt, dass es unter den aktuellen Ausbeutungsbedingungen keine Garantie dafür gibt, dass eine Versorgung auf existierendem Niveau aufrechterhalten werden kann. Der amerikanische Mahagonibaum wächst zwischen 40 und 50 m hoch und erreicht einen Stammdurchmesser von etwa 180 cm. Sein Holz sticht sowohl durch seine ästhetischen Qualitäten, seine glänzende Textur in der Farbe von Rotwein (so besonders, dass es der Mahagoni-Farbe seinen Namen gibt) und seine ruhigen Muster als auch durch seine physischen Eigenschaften hervor, wie seine Langlebigkeit, Widerstandsfähigkeit und außergewöhnliche Formbarkeit.

Scientific / Botanical name
Swietenia macrophylla

Trade / Common name
Big leaf mahogany

Family name
Meliaceae

Regions / Countries of distribution
Central and South America, and the Caribbean

Global threat status
VU

Common uses
Boats, cabinetry, furniture, musical and scientific instruments, construction materials, carvings, flooring, moldings, carpentry, veneers, interior paneling, etc.

Allgemeine Verwendung
Schiffe, Tischlerarbeiten, Mobiliar, Musikinstrumente, wissenschaftliche Instrumente, Konstruktionsmaterial, Schnitzereien, Böden, Leisten, Schreinerarbeiten, Verkleidungen, Polstermöbel, etc.

Machinability

Boring	
Carving	
Gluing	
Mortising	
Moulding	
Nailing	
Planing	
Routing and recessing	
Turning	

Physical properties

Numerical data	Green	Dry	English / Metric
Bending strength	8,844 // 621	11,514 // 809	psi // kg/cm²
Density		36 // 576	lbs/ft³ // kg/m³
Hardness		801 // 363	lbs // kg
Impact strength	30 // 76	23 // 58	in // cm
Maximum crushing strength	4,425 // 311	6,465 // 454	psi // kg/cm²
Shearing strength		1,230 // 86	psi // kg/cm²
Stiffness	1,305 // 91	1,426 // 100	1,000 psi // 1,000 kg/cm²
Weight	47 // 752	31 // 496	lbs/ft³ // kg/m³
Radial shrinkage		3	%
Tangential shrinkage		5	%

ACER MACROPHYLLUM

Description

Of all of the different maples, the *Acer macrophyllum*, big leaf maple or Oregon maple has the largest leaves: between 15 and 30 cm (6-12 in) wide and 60 cm (24 in) long. At maturity, it reaches heights of between 15 and 30 m (49-98 ft) and diameters of between 30 and 80 cm (12-31 in). Its trunk is well formed and clear up to about two-thirds of its height. The heartwood and sapwood, indistinguishable from each other, feature tones that range from saffron-colored to earthy. The wood does not respond to preservatives, which, combined with its poor natural resistance, leads its use in harsh environments to be discouraged.

Beschreibung

Unter allen Ahornarten ist der *Acer macrophyllum* oder Oregon-Ahorn derjenige mit den größten Blättern: zwischen 15 und 30 cm in der Breite und 60 cm in der Länge. Er erreicht eine Höhe von 15-30 m und einen Stammdurchmesser zwischen 30 und 80 cm. Sein Stamm ist formschön und bis auf etwa zwei Drittel seiner Höhe astrein. Splint- und Kernholz sind undifferenziert in Erd- und Safrantönen gehalten. Aufgrund seiner geringen natürlichen Widerstandsfähigkeit in Verbindung damit, dass es Schutzanstriche schlecht aufnimmt, ist von einem Einsatz des Holzes in aggressiver Umgebung abzuraten.

Scientific / Botanical name
Acer macrophyllum

Trade / Common name
Big leaf maple

Family name
Sapindaceae

Regions / Countries of distribution
Western North America

Global threat status
NE

Common uses

Baseboards, furniture, veneer, frames, plywood, pallets, surfaces, construction materials, etc.

Allgemeine Verwendung

Sockelleisten, Mobiliar, Furniere, Rahmen, Sperrholz, Paletten, Verkleidungen, Konstruktionsmaterial, etc.

Machinability

- Boring
- Gluing
- Mortising
- Moulding
- Nailing
- Planing
- Polishing
- Sanding
- Steam bending
- Turning
- Veneering qualities

Physical properties

Numerical data	Green	Dry	English / Metric
Bending strength	7,400 // 520	10,700 // 752	psi // kgf/cm^2
Hardness		850 // 385	lbs // kg
Impact strength	23 // 58	28 // 71	in // cm
Maximum crushing strength	3,240 // 227	5,950 // 418	psi // kgf/cm^2
Shearing strength		1,730 // 121	psi // kgf/cm^2
Stiffness	1,100 // 77	1,450 // 101	1,000 psi // 1,000 kgf/cm^2
Weight		34 // 544	lbs/ft^3 // kg/m^3
Radial shrinkage		4	%
Tangential shrinkage		7	%

ACER RUBRUM

Description

The American red maple or Canadian maple can tolerate soils over a wide pH range and is therefore a highly adaptive species. It is the American maple that has the most variable characteristics, though it is easily identifiable by its leaves, which measure between 5 and 10 cm (2-4 in) and have irregular dentate lobes. It is often hybridized with the silver maple, resulting in the *Acer freemanii*. The mature tree measures between 18 and 27 m (59-89 ft), and its trunk has a diameter of about 80 cm (31 in). It grows extraordinarily quickly in its first 20 to 30 years. It is less expensive than the sugar maple, which is the only maple with a greater share of the commercial market.

Beschreibung

Der Rot-Ahorn toleriert einen weiten pH-Bereich und stellt daher eine enorm anpassungsfähige Spezies dar. Wenngleich er von allen amerikanischen Ahornen die variabelsten Merkmale aufweist, lässt er sich einfach anhand seiner 5 bis 10 cm großen Blätter mit ungleichmäßig gezähnten Lappen erkennen. Er tritt regelmäßig in Mischform mit dem Silber-Ahorn auf, was als Kreuzung den *Acer freemanii* hervorbringt. Der ausgereifte Baum ist zwischen 18 und 27 m hoch und sein Stamm hat einen Durchmesser von ca. 80 cm. Der Rot-Ahorn wächst in seinen ersten 20-30 Lebensjahren außergewöhnlich schnell. Er ist preislich günstiger als der Zucker-Ahorn und nach diesem Nummer zwei der am meisten kommerziell ausgebeuteten Ahornspezies.

Scientific / Botanical name
Acer rubrum

Trade / Common name
Red maple

Family name
Sapindaceae

Regions / Countries of distribution
Eastern North America

Global threat status
NE

Common uses
Packaging, stationery, cellulose pulp, pallets, veneers, interior construction, barrels, etc.

Allgemeine Verwendung
Pakete, Verpackungen, Papierwaren, Zellstoff, Paletten, Verkleidungen, Innenausbau, Fässer, etc.

Machinability

Boring	
Gluing	
Mortising	
Nailing	
Planing	
Sanding	
Steam bending	
Turning	

Physical properties

Numerical data	Green	Dry	English / Metric
Bending strength	7,700 // 541	13,400 // 942	psi // kgf/cm2
Density		40 // 640	lbs/ft³ // kg/m³
Hardness		950 // 430	lbs // kg
Impact strength	32 // 81	32 // 81	in // cm
Maximum crushing strength	3,280 // 230	6,541 // 459	psi // kgf/cm²
Shearing strength		1,850 // 130	psi // kgf/cm²
Stiffness	1,390 // 97	1,640 // 115	1,000 psi // 1,000 kgf/cm²
Radial shrinkage		2	%
Tangential shrinkage		8	%

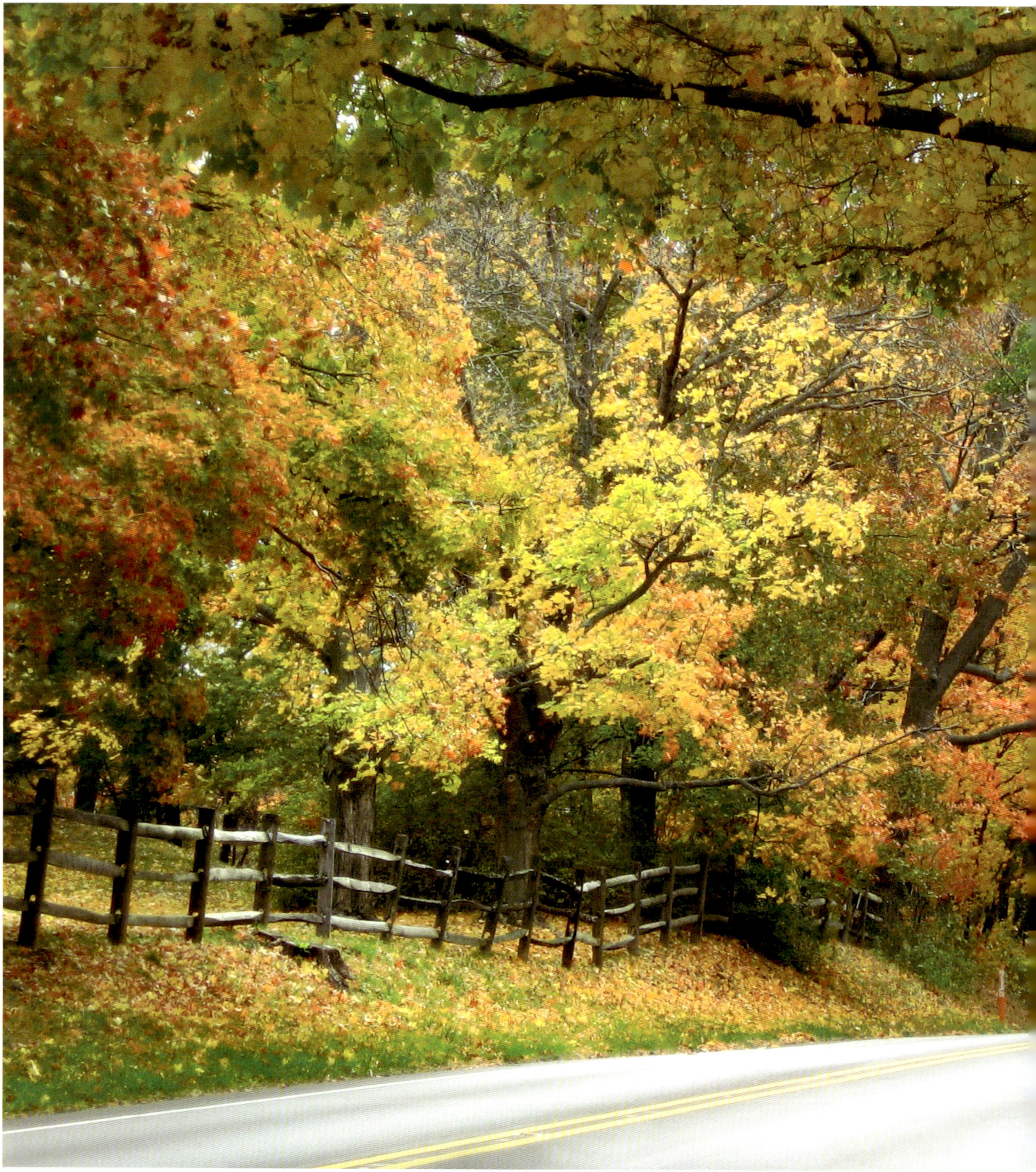

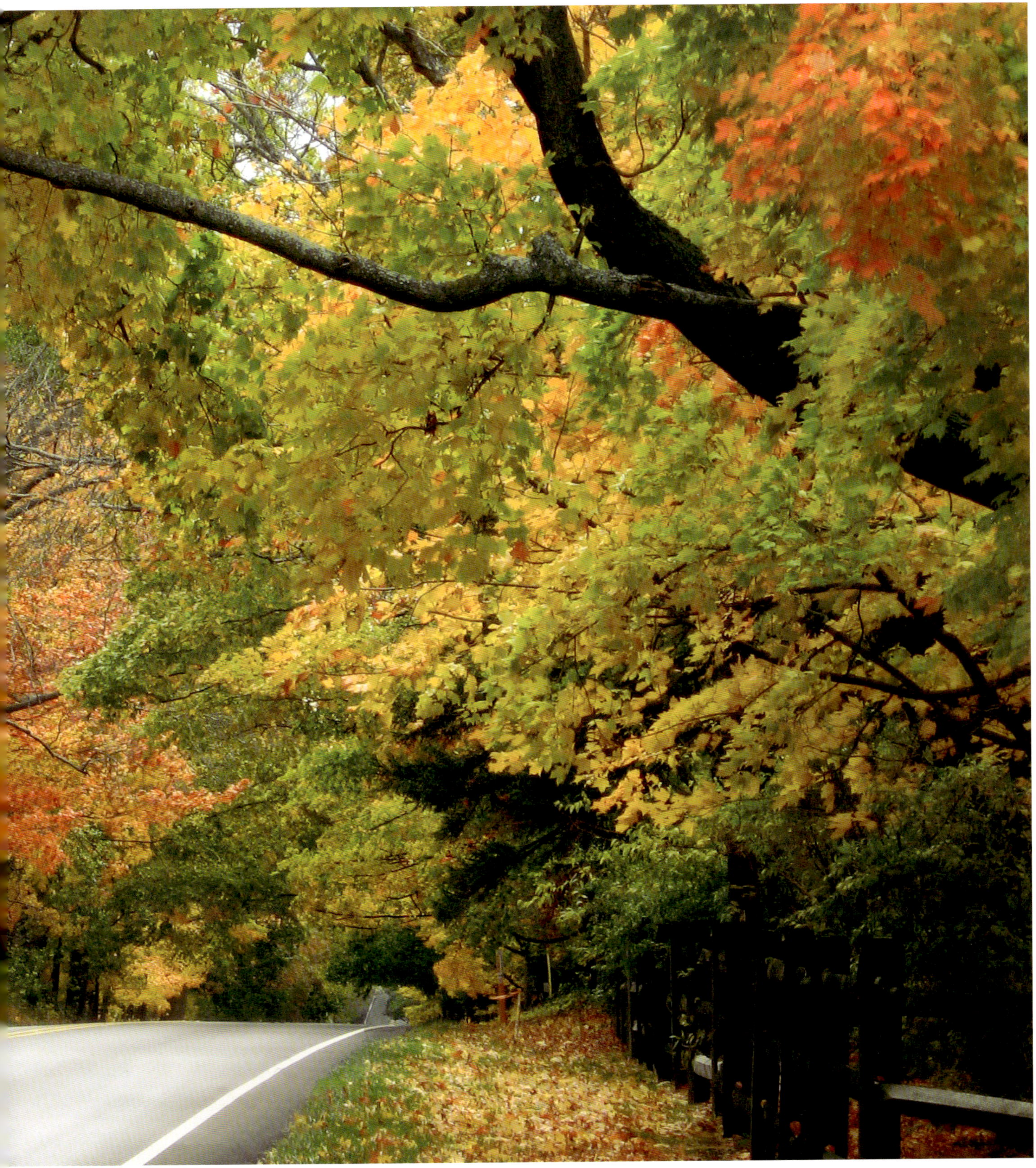

FLINDERSIA BRAYLEYANA

Description

Flindersia brayleyana does not exceed 40 m (131 ft) in height. The trunk, which can reach 2.5 m (8 ft) in diameter, is circular in cross-section and not buttressed. With a spreading crown and large clusters of off-white flowers, Queensland maple has a majestic growth habit that does not record age with growth rings. The bark is dark grayish brown with distinct longitudinal fissures. Leaves are pinnate and ovate-elliptic. The sapwood of the Queensland maple is pale gray and the heartwood is pink to brownish pink making for a distinctive appearance that finishes well.

Beschreibung

Flindersia brayleyana erreicht eine Höhe von bis zu 40 m. Der Stamm kann durchschnittlich bis zu 2,5 m wachsen, hat einen runden Querschnitt und hat keine breit auslaufende Basis. Mit seiner ausladenden Krone und den großen Zapfen von grauweißen Blüten gehört der Queensland Ahorn zu den schnellwachsenden Bäumen, dessen Alter nicht anhand von Wachstumsringen bestimmt werden kann. Die Rinde ist dunkelgraubraun mit ausgeprägten Längsrissen. Die Blätter sind gefiedert und eiförmig-elliptisch. Das Splintholz des Queensland Ahorns ist hellgrau und das Kernholz ist pink bis bräunlich pink, was für eine markantes Erscheinungsbild sorgt und gut verarbeitet werden kann.

Scientific / Botanical name
Flindersia brayleyana

Trade / Common name
Queensland maple

Family name
Rutaceae

Regions / Countries of distribution
Northern Queensland in Australia

Global threat status
NE

Common uses
The timber of the Queensland maple is used for cabinets and indoor fittings thanks to its attractive pink tones and wavy grain but it is hard to come by since the areas in which this tree grows are protected.

Allgemeine Verwendung
Dank seiner attraktiven pinken Farbgebung und der welligen Maserung wird das Schnittholz des Queensland Ahorns für die Herstellung von Schränken und Inneneinbauten eingesetzt. Da der Baum in geschützten Gebieten wächst, ist es schwierig die Nachfrage zu decken.

Machinability

Boring	
Gluing	
Mortising	
Moulding	
Painting	
Planing	
Polishing	
Staining	
Turning	

Physical properties

Numerical data	Green	Dry	English / Metric
Bending strength	7,180 // 504	11,250 // 790	psi // kgf/cm^2
Density		35 // 560	lbs/ft^3 // kg/m^3
Hardness		990 // 449	lbs // kg
Maximum crushing strength	3,915 // 275	6,400 // 449	psi // kgf/cm^2
Shearing strength		1,880 // 132	psi // kgf/cm^2
Stiffness	1,300 // 91	1,505 // 105	1,000 psi // 1,000 kgf/cm^2
Weight	32 // 512	25 // 400	lbs/ft^3 // kg/m^3
Radial shrinkage		5	%
Tangential shrinkage		9	%

SCLEROCARYA BIRREA

Description

Marula is an African deciduous tree belonging to the same plant family as mangos and cashews. It grows up to 18 m (59 ft) and develops an erect trunk and a rounded symmetrical crown. The bark is mottled and flakes in roundish patches exposing the light yellow inner bark. Leaves, clustered at the end of branches, are composite, forming 2 to 23 leaflets that are shiny dark green on top and light green underneath. The pink and white flowers of the marula are male and female on separate trees. Male flowers form long droopy sprays and female flowers bloom singly or in small clusters. The fleshy fruits, yellow and egg-shaped are highly prized by humans and animals. Wood is light reddish brown to whitish with no definite heartwood.

Beschreibung

Der Marula-Baum ist ein sommergrüner Baum aus Afrika, welcher der selben Pflanzenfamilie angehört wie Mangos und Cashews. Er erreicht eine Höhe von bis zu 18 m, hat einen aufrechten Stamm und eine symmetrisch abgerundete Krone. Die Rinde ist gefleckt und blättert in rundlichen Platten ab, unter denen die hellgelbe Innenrinde zum Vorschein kommt. Seine Blätter, die auf der Oberseite dunkelgrün glänzend und an der Unterseite hellgrün sind, wachsen in Büscheln von 2-23 Blättchen zusammengefasst am Ende eines Astes. Die pinken und weißen Blüten des Marula-Baums sind männlich und weiblich und kommen an separaten Bäumen vor. Die männlichen Blüten bilden lange, herabhängende Sträuße, während die weiblichen Blüten entweder einzeln oder in kleinen Verbünden blühen. Die fleischigen Früchte des Marula-Baums sind gelb und eiförmig und bei Mensch und Tier sehr beliebt. Die Farbe seines Holzes variiert zwischen einem hellen Rotbraun bis hin ins Weißliche, wobei das Kernholz nicht klar umrissen ist.

Scientific / Botanical name
Sclerocarya birrea

Trade / Common name
Marula

Family name
Anacardiaceae

Regions / Countries of distribution
Southern Africa

Global threat status
NE

Common uses

The wood of the marula is used for furniture, paneling, flooring and carving. Rope is made from the inner bark and a red-brown dye can be extracted from the fresh part of the bark.

Allgemeine Verwendung

Verwendet wird das Holz bei der Herstellung von Mobiliar, für Holzvertäfelungen, als Bodenbelag und für Schnitzarbeiten. Aus der Innenrinde des Marula-Baums werden Taue hergestellt und aus dem frischen Teil der Rinde kann rotbraune Farbe gewonnen werden.

Machinability

Boring	
Gluing	
Mortising	
Moulding	
Nailing	
Planing	
Turning	
Veneering qualities	

Physical properties

Numerical data	Green	Dry	English / Metric
Bending strength	4,510 // 317	6,940 // 487	psi // kgf/cm^2
Density		30 // 480	lbs/ft^3 // kg/m^3
Hardness		995 // 451	lbs // kg
Maximum crushing strength	2,140 // 150	3,915 // 275	psi // kgf/cm^2
Shearing strength		1,270 // 89	psi // kgf/cm^2
Stiffness	1,020 // 71	1,200 // 84	1,000 psi // 1,000 kgf/cm^2
Weight	30 // 480	26 // 416	lbs/ft^3 // kg/m^3
Radial shrinkage		5	%
Tangential shrinkage		9	%

MANGIFERA INDICA

Description

This is a species of mango tree native to India, the cultivated varieties of which have been introduced in other temperate regions around the world. This is the largest fruit tree in existence, capable of reaching heights of between 24 and 30 m (80-100 ft) and diameters of between 90 and 120 cm (36-48 in). Its blackish bark emits a resinous latex. Its wood is moderately hard, with a lustrous texture, slightly rough and very porous, with clearly visible growth circles, straight grain, a sapwood with creamy or brown tones and a heartwood with cinnamon tones and black streaks. It is a fairly malleable and termite-resistant material.

Beschreibung

Die Mango ist eine Pflanzenart, die ursprünglich aus Indien stammt, heute jedoch als Kulturpflanze in weiten Teilen der Welt mit gemäßigtem Klima vorkommt. Mit einer Höhe von 24 bis 30 m und einem Stammdurchmesser von etwa 90 bis 120 cm ist die Mango der größte Obstbaum, den es gibt. Die schwarzbraune Rinde sondert Latex ab. Das Holz der Mango ist mäßig hart, von glänzender Textur, leicht grob und sehr porös. Wachstumsringe sind deutlich erkennbar und die Maserung ist geradlinig. Das Splintholz ist in Creme- oder Brauntönen gehalten und das Kernholz hat die Farbe von Zimt und ist teilweise mit schwarzen Adern durchzogen. Es handelt sich um ein gut formbares Holz, dass gegen Termiten resistent ist.

Scientific / Botanical name
Mangifera indica

Trade / Common name
Mango

Family name
Anacardiaceae

Regions / Countries of distribution
India

Global threat status
DD

Common uses

Although its agricultural use far exceeds its industrial use, the mango tree is used in south Asia for the production of submerged pilings, boats, furniture, carpentry, flooring, packaging and crates.

Allgemeine Verwendung

Wenngleich die landwirtschaftliche Nutzung des Mangobaumes die industrielle Nutzung stark übersteigt, wird sein Holz im Süden Asiens für die Herstellung von Pfosten, Booten, Mobiliar, Schreinerarbeiten, Böden, Verpackungen und Kisten eingesetzt.

Machinability

Boring	
Mortising	
Moulding	
Planing	
Polishing	
Sanding	
Turning	

Physical properties

Numerical data	Green	Dry	English / Metric
Bending strength	7,799 // 548	13,790 // 969	psi // kg/cm^2
Density		40 // 640	lbs/ft^3 // kg/m^3
Hardness		1,000 // 453	lbs // kg
Impact strength	27 // 68		in // cm
Maximum crushing strength	3,755 // 264	6,887 // 484	psi // kg/cm^2
Stiffness	1,161 // 81	1,745 // 122	1,000 psi // 1,000 kg/cm^2
Weight	55 // 881	43 // 688	lbs/ft^3 // kg/m^3
Radial shrinkage		4	%
Tangential shrinkage		5	%

MELIA AZEDARACH

Description

A member of the Meliaceae family, the beadtree is a deciduous tree species native to Southeast Asia and introduced in South Africa and the Americas, where it is an invasive species that has displaced indigenous species. Bearing extremely toxic fruit, it reproduces and grows very easily, reaching heights of between 8 and 15 m (26-39 ft) and diameters of about 30 cm (12 in). It produces a wood with a rough and irregular texture, with generally straight grain, though sometimes interwoven, and a yellowish sapwood very different from the heartwood, which has cinnamon or dark brown tones. It is a high-quality material though it is underused commercially.

Beschreibung

Der Zedrachbaum, auch Persischer Flieder, Chinesischer Holunder oder Paternosterbaum genannt, ist Mitglied der Familie der Meliaceae. Er ist eine laubabwerfende Spezies, die aus dem Nordosten Asiens stammt und in Südafrika und Amerika verbreitet ist, wo er zur invasiven Art wurde und heimische Arten verdrängte. Die Früchte des Zedrachbaumes sind äußerst giftig. Er verbreitet sich und wächst mit wunderbarer Leichtigkeit, wobei er Höhen zwischen 8 und 15 m erreicht. Sein Stammdurchmesser beträgt in etwa 30 cm. Sein Holz ist von rauer und unregelmäßiger Textur mit einer Maserung, die zumeist geradlinig, jedoch ab und zu auch verflochten ist. Der gelbliche Splint setzt sich klar vom Kernholz ab, das in Zimt- und dunklen Kastanientönen gehalten ist. Es handelt sich um hochwertiges Material, das jedoch vom kommerziellen Standpunkt aus zu wenig genutzt wird.

Scientific / Botanical name
Melia azedarach

Trade / Common name
Bead-tree

Family name
Meliaceae

Regions / Countries of distribution
Pakistan, India, Southeast Asia, and Australia

Global threat status
NE

Common uses

Although it is difficult to obtain given its residual commercial distribution, its wood can be used in the production of crates and packaging, furniture, plywood and veneers, roofing, flooring, etc.

Allgemeine Verwendung

Obwohl das Holz des Zedrachbaumes aufgrund des geringen kommerziellen Vertriebs nur schwer erhältlich ist, könnte es für die Herstellung von Kisten und Verpackungen, Mobiliar, Sperrholz und Verkleidungen von Nutzen sein.

Machinability

Boring	
Gluing	
Mortising	
Moulding	
Nailing	
Planing	
Polishing	
Sanding	
Staining	
Turning	

Physical properties

Numerical data	Green	Dry	English / Metric
Bending strength	8,520 // 599	11,850 // 833	psi // kg/cm²
Density		36 // 576	lbs/ft³ // kg/m³
Hardness		1,027 // 465	lbs // kg
Impact strength	51 // 129	46 // 116	in // cm
Maximum crushing strength	3,925 // 275	7,000 // 492	psi // kg/cm²
Shearing strength		2,042 // 143	psi // kg/cm²
Stiffness	1,162 // 81	1,309 // 92	1,000 psi // 1,000 kg/cm²
Weight	50 // 800	37 // 592	lbs/ft³ // kg/m³
Radial shrinkage		5	%
Tangential shrinkage		9	%

ACER CAMPESTRE

Description

The *Acer campestre* (also known as the field maple or hedge maple) is the only maple species native to the United Kingdom. It has properties similar to the sycamore, though its wood is harder. Its small size, about 20 m (66 ft) at maturity, make it non-viable as a commercial wood. The wood initially has a creamy white color, which becomes the color of cinnamon with time. The heartwood and sapwood are indistinguishable from each other. The wood is especially vulnerable to fungi and insects. Its texture is smooth on the surface and its appearance is relatively homogeneous. The sapwood responds well to preservatives, but this is not the case for the heartwood.

Beschreibung

Der *Acer campestre*, auch als Feldahorn oder Maßholder bekannt, ist die einzige Spezies der endemischen Ahorne des Vereinigten Königreichs. Er besitzt Eigenschaften ähnlich denen der Sykomore, nur dass sein Holz härter ist. Einmal ausgereift beträgt seine geringe Höhe etwa 20 m, was ihn für die kommerzielle Verwendung unbrauchbar macht. Ursprünglich ist die Farbe des Holzes ein Cremeweiß, das sich jedoch mit der Zeit in einen Zimtton ändert. Splint- und Kernholz lassen sich nicht unterscheiden. Das Holz ist extrem pilz- und insektenanfällig. Seine Oberflächentextur ist geschmeidig, das Erscheinungsbild recht homogen. Im Gegensatz zum Kernholz nimmt das Splintholz Schutzanstriche gut auf.

Scientific / Botanical name
Acer campestre

Trade / Common name
Field maple

Family name
Sapindaceae

Regions / Countries of distribution
Europe, southwest Asia, north of Africa, North America, and Australia

Global threat status
NE

Common uses

This species is highly valued in cosmetics, apiculture, carpentry and cabinetry.

Allgemeine Verwendung

Kosmetik, Imkerei, Tischler- und Schreinerhandwerk sind Bereiche, in denen die Spezies sehr geschätzt wird.

Machinability

Gluing	
Moulding	
Nailing	
Planing	
Polishing	
Staining	
Steam bending	
Turning	

Physical properties

Numerical data	Green	Dry	English / Metric
Bending strength		15,800 // 1,110	psi // kgf/cm^2
Hardness		1,150 // 521	lbs // kg
Weight		44 // 704	lbs/ft^3 // kg/m^3
Radial shrinkage		3	%
Tangential shrinkage		8	%

SWIETENIA MAHAGONI

Description

While the Honduras mahogany is the most exploited today, our next species, the West Indies mahogany, was the first species to be used in Europe in the 16th century, where the famous mahogany wood was given its name. Overexploited for more than two centuries —especially as a result of the enormous boom in the shipbuilding industry during the colonial period— the species is currently endangered and its trade is therefore almost token in nature. It remains, however, a popular ornamental tree in parks and along avenues in tropical and subtropical regions around the world. At maturity, the species reaches heights of more than 46 m (150 ft) and diameters of about 180 cm (60-80 ft).

Beschreibung

Wenngleich das Amerikanische Mahagoni heutzutage am häufigsten vertrieben wird, ist das Swientenia Mahagoni, die Spezies, die im XVI. Jahrhundert als erste in Europa eingeschifft wurde und die dem wohlbekannten Mahagoniholz seinen Namen verliehen hat. Speziell aufgrund des brutalen Aufschwungs der Schiffsindustrie im Zuge des Kolonialismus, wurde mehr als zwei Jahrhunderte lang Raubbau an der Spezies getrieben. Aktuell ist sie bedroht und ihr Vertrieb ist quasi Zeugnis gebend. Ungeachtet dessen bleib es ein beliebter Zierbaum, der häufig in den Parkanlagen und an den Prachtstraßen der tropischen- und subtropischen Gegenden auf der Welt zu finden ist. Ausgewachsen erreicht die Spezies eine Höhe von über 46 m und einen Stammdurchmesser von etwa 180 cm.

Scientific / Botanical name
Swietenia mahagoni

Trade / Common name
West Indies mahogany

Family name
Meliaceae

Regions / Countries of distribution
Southern eastern United States, the Bahamas, Cuba, St. Croix, Virgin Islands, and Hispaniola

Global threat status
EN

Common uses

Home flooring, crates, baskets, trunks and containers, veneers, planks, construction materials, cabinetry, pallets, carpentry, boats, etc.

Allgemeine Verwendung

Böden für den Wohnbereich, Kisten, Körbe, Truhen und Behälter, Verkleidungen, Arbeitsplatten, Konstruktionsmaterial, Tischlerarbeiten, Paletten, Schreinerarbeiten, Schiffbau, etc.

Machinability

Boring	
Carving	
Gluing	
Mortising	
Moulding	
Nailing	
Planing	
Routing and recessing	
Sanding	
Turning	

Physical properties

Numerical data	Green	Dry	English / Metric
Bending strength		12,782 // 898	psi // kg/cm^2
Maximum crushing strength		7,192 // 505	psi // kg/cm^2
Stiffness		1,285 // 90	1,000 psi // 1,000 kg/cm^2
Weight	47 // 752	35 // 560	lbs/ft^3 // kg/m^3
Radial shrinkage		3.7	%
Tangential shrinkage		4.4	%

CARAPA GUIANENSIS

Description

This is a large tree that reaches its greatest height —up to 52 m (170 ft)— in the Amazon basin, with diameters of between 150 and 180 cm (60-72 in) at the trunk. Its fruit is a nut similar to a chestnut. The sapwood is rosy but becomes cinnamon or gray in color after exposure. The heartwood has a salmon color which becomes more subdued after oxidation. The texture of the wood is very grainy, with somewhat irregular parallel stripes and a slightly rough surface. It is an important commercial wood and a fundamental construction material in its natural areas of distribution. Its physical properties make it a potential substitute for mahogany.

Beschreibung

Baum mit großen Dimensionen, der seine Maximalhöhe von bis zu 52 m im Amazonasbecken erreicht. Sein Stammdurchmesser kann zwischen 150 und 180 cm betragen. Seine Frucht ist eine Nuss, die an die Kastanie erinnert. Der Splint ist rosenrot, wird jedoch zimtfarben oder gräulich, wenn er dem Licht ausgesetzt ist. Das Kernholz ist lachsfarben, verblasst jedoch durch Oxidation. Die Textur des Holzes ist sehr buntäderig, mit teils unregelmäßigen parallelen Linien und einer leicht groben Oberfläche. Es handelt sich um ein Holz mit großer kommerzieller Bedeutung, das in seinem natürlichen Verbreitungsgebiet als Konstruktionsmaterial unverzichtbar ist. Seine physischen Eigenschaften machen es zu einem potentiellen Ersatz für die Caboas.

Scientific / Botanical name
Carapa guianensis

Trade / Common name
Najesí

Family name
Meliaceae

Regions / Countries of distribution
Tropical South America and Africa

Global threat status
NE

Common uses

The oil from its seeds is used for drugs, soaps, insecticides and margarine. Its wood is found in carpentry, furniture, construction, plywood, flooring, etc.

Allgemeine Verwendung

Aus dem Öl seiner Samen werden Medikamente, Seifen, Insektizide und Margarine hergestellt. Sein Holz wird für Schreinerarbeiten, für die Herstellung von Mobiliar, für Konstruktionen, als Pressholz, für Böden, etc. verwendet.

Machinability

Boring	
Carving	
Gluing	
Mortising	
Moulding	
Nailing	
Planing	
Polishing	
Routing and recessing	
Sanding	
Staining	
Turning	
Varnishing	

Physical properties

Numerical data	Green	Dry	English / Metric
Bending strength	10,903 // 766	15,583 // 1,095	psi // kgf/cm^2
Hardness		1,220 // 553	lbs // kg
Maximum crushing strength	5,096 // 358	8,281 // 582	psi // kgf/cm^2
Shearing strength		1,495 // 105	psi // kgf/cm^2
Stiffness	1,690 // 118	2,000 // 140	1,000 psi // 1,000 kgf/cm^2
Weight	59 // 945	40 // 640	lbs/ft^3 // kg/m^3
Radial shrinkage		4	%
Tangential shrinkage		8	%

KHAYA SENEGALENSIS

Description

Khaya senegalensis is a deciduous evergreen tree that grows naturally in forests of West Africa but is threatened by habitat loss. It reaches 10 to 15 m (32 to 49 ft) in height and a trunk diameter of 1 m (3 ft). The pink-brown heartwood is slightly darker than the sapwood and not clearly delineated. Compared to other varieties of the same species, *Khaya senegalensis* has prominent interlocked grain that produces a striped figure.

Beschreibung

Khaya senegalensis ist ein sommergrüner Baum, der normalerweise in den Wäldern Westafrikas zuhause, jedoch vom Verlust seines Lebensraums bedroht ist. Bei einem Stammumfang von 1 m erreicht er eine Höhe von 10 bis 15 m. Das pink-braune Kernholz ist wenig dunkler als das Splintholz und nicht klar von diesem abgegrenzt. Verglichen mit anderen Arten aus der selben Familie, hat *Khaya senegalensis* eine markante, verzahnte Faserstruktur, die eine gestreifte Maserung hervorbringt.

Scientific / Botanical name
Khaya senegalensis

Trade / Common name
African mahogany

Family name
Meliaceae

Regions / Countries of distribution
Central Africa

Global threat status
VU

Common uses

Because of its decorative quality, the timber is favored for furniture, high-end joinery, trim, boat construction, flooring, turnery and veneers.

Allgemeine Verwendung

Aufgrund seiner dekorativen Qualität wird das Schnittholz bevorzugt für Mobiliar, hochwertige Schreinerarbeiten, Innenausstattung, Bootsbau, Bodenbelag, Drechslerarbeiten und als Furnier verwendet.

Machinability

Boring	
Carving	
Gluing	
Mortising	
Moulding	
Nailing	
Planing	
Polishing	
Routing and recessing	
Sanding	
Turning	
Varnishing	

Physical properties

Numerical data	Green	Dry	English / Metric
Bending strength	8,250 // 580	12,550 // 882	psi // kgf/cm²
Density		46 // 736	lbs/ft³ // kg/m³
Hardness		1,350 // 612	lbs // kg
Maximum crushing strength	4,125 // 290	7,225 // 507	psi // kgf/cm²
Shearing strength		1,800 // 126	psi // kgf/cm²
Stiffness	1,435 // 100	1,675 // 117	1,000 psi // 1,000 kgf/cm²
Weight		50 // 800	lbs/ft³ // kg/m³
Radial shrinkage		3.7	%
Tangential shrinkage		6.6	%

ACER PALMATUM

Description

Typically known as the Japanese maple, *Acer palmatum* is a small tree or shrub that reaches heights of 6-10 m (20-33 ft) and diameters of around 30 cm (12 in). Its fruit consists of a pair of samaras measuring 2 to 3 cm (0.8-1.2 in). Its palmate-lobed leaves exhibit an incredible variety of colors throughout the seasons, which makes them highly valued as decorative objects. It is a dense and heavy wood that is used as an alternative to the sugar maple (*Acer saccharum*) in certain applications. The sapwood and heartwood share cinnamon and brownish-pink colors.

Beschreibung

Der *Acer palmatum* ist üblicherweise als Fächerahorn bekannt. Entweder ist er ein kleiner Baum oder ein Busch, der eine Höhe von 6-10 m und einen Stammdurchmesser von etwa 30 cm erreicht. Er trägt eine Flügelfrucht von 2 bis 3 cm Größe. Seine handförmigen, lappigen Blätter durchlaufen im Zuge der Jahreszeiten eine breite Farbpalette und sind daher beliebtes Dekorationsobjekt. Das Holz des Fächerahorns ist dicht und schwer und wird in bestimmten Bereichen alternativ zum Holz des Zucker-Ahorns (*Acer saccharum*) eingesetzt. Splint- und Kernholz sind zimtfarben und kastanienrot.

Scientific / Botanical name
Acer palmatum

Trade / Common name
Japanese maple

Family name
Sapindaceae

Regions / Countries of distribution
Japan, North Korea, South Korea, China, eastern Mongolia, and southeast Russia

Global threat status
NE

Common uses

Given its famous resistance to wear, it is ideal for flooring. Other uses include textiles, spools, bolts, veneer and plywood, surfaces, paneling, etc.

Allgemeine Verwendung

Aufgrund seiner allgemein bekannten Strapazierfähigkeit eignet sich das Holz hervorragend als Bodenbelag. Zudem wird es für die Herstellung von Textilien, Spulen, Bolzen, Furniere und Sperrholz, Verkleidungen, Polstermöbel, etc.

Machinability

Gluing	
Nailing	
Planing	
Polishing	
Staining	
Steam bending	

Physical properties

Numerical data	Green	Dry	English / Metric
Bending strength	9,400 // 660	15,800 // 1110	psi // kgf/cm^2
Density		40 // 640	lbs/ft^3 // kg/m^3
Hardness		1,450 // 657	lbs // kg
Maximum crushing strength	4,020 // 282	7,830 // 550	psi // kgf/cm^2
Shearing strength		2,330 // 163	psi // kgf/cm^2
Stiffness	1,550 // 108	1,830 // 128	1,000 psi // 1,000 kgf/cm^2
Weight	52 // 832	40 // 640	lbs/ft^3 // kg/m^3
Radial shrinkage		5	%
Tangential shrinkage		10	%

ACER SACCHARUM

Description

The *Acer saccharum* or sugar maple is a wide, columnar species with palmate-lobed leaves. Its name comes from one of its by-products, maple sugar. Its bark is grayish-brown and smooth, and becomes cracked and flaky with age. It grows to heights of 21 to 37 m (69-121 ft), with diameters of 60 to 90 cm (24-35 in). It has small, greenish-yellow, hanging flowers. The heartwood has a uniform reddish or cinnamon color. The sapwood exhibits white tones with reddish tints. It is the most commercially exploited maple species. It is an excellent tree for use in parks and gardens because it is easy to transplant and propagate and it grows rapidly.

Beschreibung

Der *Acer saccharum* oder Zucker-Ahorn ist eine breite und säulenartige Spezies mit handförmigen, lappigen Blättern. Er verdankt seinen Namen einem seiner Nebenerzeugnisse, dem Ahornzucker. Seine Rinde ist graubraun und glatt und wird mit zunehmendem Alter rissig und schuppig. Er wächst zwischen 21 und 37 m hoch und hat einen Stammdurchmesser zwischen 60 und 90 cm. Seine hängenden Blüten sind klein und gelbgrün. Das Kernholz ist entweder rötlich oder einheitlich zimtfarben. Das Splintholz zeigt sich in Weißtönen mit rötlichen Nuancen. Der Zucker-Ahorn ist die am meisten kommerziell vertriebene Spezies unter den Ahornen. Weil er sich einfach umpflanzen und vermehren lässt und schnell wächst, eignet er sich hervorragend für die Bepflanzung von Parkanlagen und Gärten.

Scientific / Botanical name
Acer saccharum

Trade / Common name
Sugar maple

Family name
Sapindaceae

Regions / Countries of distribution
Northeastern North America

Global threat status
NE

Common uses
Shipbuilding, packaging, furniture, carpentry, veneers, moldings, paneling, interior construction, etc.

Allgemeine Verwendung
Boots- und Schiffskonstruktion, Verpackungen und Pakete, Mobiliar, Schreinerarbeiten, Verkleidungen, Leisten, Polstermöbel, Innenausbau, etc.

Machinability

Boring	
Carving	
Gluing	
Mortising	
Moulding	
Nailing	
Planing	
Sanding	
Turning	
Veneering qualities	

Physical properties

Numerical data	Green	Dry	English / Metric
Bending strength	9,400 // 660	15,800 // 1,110	psi // kgf/cm²
Hardness		1,450 // 657	lbs // kg
Maximum crushing strength	4,020 // 282	7,830 // 550	psi // kgf/cm²
Shearing strength		2,330 // 163	psi // kgf/cm²
Stiffness	1,550 // 108	1,830 // 128	1,000 psi // 1,000 kgf/cm²
Weight	56 // 897	44 // 704	lbs/ft³ // kg/m³
Radial shrinkage		5	%
Tangential shrinkage		10	%

CALODENDRUM CAPENSE

Description

The common name of this species, the Cape chestnut, was the result of a misunderstanding: the explorer William Burchell confused its pink flowers with those of the chestnut, a species to which it has no relation, as it is a citrus tree, a member of the Rutaceae family. In forests, it grows to heights of 20 m (66 ft), but when it is found in the plains or when cultivated, it rarely surpasses 10 m (33 ft). The trunk is smooth and a soft gray color, with the oldest specimens dotted by lichens. Its wood exhibits cinnamon-colored tones; it is hard yet flexible, and thus fairly malleable. It is a slow-growing, late-blooming tree common in urban areas in southern Africa.

Beschreibung

Seinen allgemeinsprachlichen Namen, Kapkastanie, verdankt der Baum einem Missverständnis: Der Entdecker William Burchell verwechselte seine rosenroten Blüten mit denen der Kastanie. In Wirklichkeit steht die Kapkastanie in keinerlei Verbindung mit dieser Spezies, da sie zur Tribus der Zitrusfrüchte aus der Familie der Rutaceae gehört. In Wäldern erreicht er eine Höhe von 20 m, wächst er jedoch auf einer Ebene oder unter menschlicher Kultivierung, überschreitet seine Höhe nur selten die Marke von 10 m. Der Stamm ist eben und hat eine sanft gräuliche Färbung, an älteren Exemplaren mit Flechte durchsetzt. Sein Holz ist in Zimttönen gehalten. Es ist hart, jedoch gleichzeitig elastisch und daher recht formbar. Es handelt sich um einen langsamwachsenden Baum mit später Blüte, der vor allem im urbanen Raum des südlichen Afrikas zu finden ist.

Scientific / Botanical name
Calodendrum capense

Trade / Common name
Cape chestnut

Family name
Rutaceae

Regions / Countries of distribution
Africa

Global threat status
NE

Common uses

The oils from its bark are used as an ingredient in cosmetics. Its seeds are used to produce soap. Its wood can be found in anvils, handles, planks, canopies and tents, etc.

Allgemeine Verwendung

Das Öl aus seiner Rinde wird im kosmetischen Bereich eingesetzt. Seine Samen dienen zur Herstellung von Seife. Sein Holz wird für die Herstellung von Ambossen, Griffen und Geländern, Arbeitsplatten, das Tragwerk von Zelten, etc verwendet.

Machinability

Boring	
Gluing	
Mortising	
Nailing	
Planing	
Polishing	
Sanding	
Steam bending	
Turning	
Varnishing	

Physical properties

Numerical data	Green	Dry	English / Metric
Bending strength	9,310 // 654	13,566 // 953	psi // kgf/cm^2
Density		42 // 672	lbs/ft^3 // kg/m^3
Hardness		1,658 // 752	lbs // kg
Maximum crushing strength	3,916 // 275	6,397 // 449	psi // kgf/cm^2
Shearing strength		2,205 // 155	psi // kgf/cm^2
Stiffness	1,460 // 102	1,686 // 118	1,000 psi // 1,000 kgf/cm^2
Weight	41 // 656	33 // 528	lbs/ft^3 // kg/m^3
Radial shrinkage		3	%
Tangential shrinkage		7	%

AILANTHUS ALTISSIMA

Description

The ailanthus or tree of heaven is a species native to northern China, though it was introduced in the 18th century in Australia, southern Europe and the United States. It grows to heights of 20 to 25 m (66-82 ft) and diameters of about a meter (39 in). Today, it is considered an invasive species a high colonization potential given its rapid growth and its ability to thrive in degraded environments. It is suitable for use in urban settings or as a landscaping tree. Its male flowers have a unique foul smell and cause allergies. Its fruits —yellowish or brownish-red samaras— are plentiful. Its wood is highly porous, weak and not very resistant, and, consequently, non-viable for commercial use.

Beschreibung

Der Götterbaum, auch Himmelsbaum oder Bitteresche genannt, ist eine Spezies, die aus dem Norden Chinas stammt, wenngleich sie im XVIII. Jahrhundert auch in Australien, Südeuropa und den Vereinigten Staaten eingeführt wurde. Der Götterbaum erreicht eine Höhe von 20 bis 25 m und einen Stammdurchmesser von etwa einem Meter. Aufgrund seines schnellen Wachstums und seiner Fähigkeit, auf wenig entwickelten Böden zu wachsen, wird er heutzutage als invasive Art mit hohem Pionierpotential betrachtet. Er eignet sich für die Pflanzung in urbaner Umgebung oder für den Einsatz im Landschaftsbau. Seine männlichen Blüten sondern einen übel riechenden Geruch ab und rufen Allergien hervor. Er ist reich an Flügelfrüchten, die von gelblicher oder rotbrauner Farbe sind. Sein poröses, instabiles und wenig widerstandsfähiges Holz ist für kommerzielle Zwecke ungeeignet.

Scientific / Botanical name
Ailanthus altissima

Trade / Common name
Ailanthus

Family name
Simaroubaceae

Regions / Countries of distribution
Central China, and Taiwan

Global threat status
NE

Common uses

It is essentially used in urban parks (especially in southern Europe), forming ornamental clusters, or for lining roads and paths.

Allgemeine Verwendung

Der Götterbaum wird hauptsächlich in städtischen Parks (vor allem im Süden Europas) eingesetzt, wo er gruppenweise als Zierpflanze steht. Häufig säumt er auch Straßen oder Wege.

Machinability

Boring	
Gluing	
Mortising	
Moulding	
Painting	
Planing	
Polishing	
Staining	
Turning	

Physical properties

Numerical data	Green	Dry	English / Metric
Weight		42 // 672	lbs/ft^3 // kg/m^3
Radial shrinkage		4.1	%
Tangential shrinkage		7.6	%

CEDRELA FISSILIS

Description

The Brazilian cedar, also known as the ygary, is a febrifuge and melliferous species that exists in colonies in the humid forests of Central America and South America. It can grow up to heights of 30 m (98 ft), with diameters of 1.5 to 2 m (39-59 in). Its wood has excellent mechanical properties: it is somewhat heavy but not very resistant, and is also very stable, incredibly malleable and practically rot-proof. It has a rather fine and lustrous texture— with golden rays, a generally straight and occasionally irregular grain and ill-defined sapwood and heartwood, the tones of which range from cinnamon to mahogany.

Beschreibung

Die Brasilianische Zeder ist eine fiebersenkende und honigbildende Spezies, die die Regenwälder Zentral- und Südamerikas besiedelt. Sie hat das Potential, bis auf eine Höhe von 30 m zu wachsen, wobei ihr Stammdurchmesser bei 1,5 bis 2 m liegt. Ihr Holz hat exzellente mechanische Eigenschaften: Es ist recht schwer, jedoch nicht besonders widerstandsfähig, sehr stabil, unglaublich formbar und praktisch unverweslich. Das Holz hat eine recht zarte und glänzende Textur – gold strahlend –, mit in der Regel geradliniger, manchmal unregelmäßiger Maserung. Splint- und Kernholz heben sich leicht voneinander ab, wobei sie farblich zwischen Zimt und Mahagoni schwanken.

Scientific / Botanical name
Cedrela fissilis

Trade / Common name
Brazilian cedar

Family name
Meliaceae

Regions / Countries of distribution
Argentina, Bolivia, Brazil, Colombia, Costa Rica, Ecuador, Panama, Paraguay, Peru, and Venezuela

Global threat status
EN

Common uses
Its poor resistance inhibits its use as a structural material. However, its durability and aesthetic appeal make it suitable for cabinetry, flooring, veneers and carpentry.

Allgemeine Verwendung
Die geringe Widerstandskraft des Holzes schließt eine Verwendung als Konstruktionsmaterial aus. Seine Langlebigkeit und attraktive Ästhetik machen es jedoch ideal für Tischlerarbeiten, Böden, Verkleidungen oder Schreinerarbeiten.

Machinability

Boring	
Carving	
Gluing	
Mortising	
Moulding	
Nailing	
Planing	
Polishing	
Screwing	
Staining	
Steam bending	
Turning	
Veneering qualities	

Physical properties

Numerical data	Green	Dry	English / Metric
Bending strength	6,111 // 429	9,589 // 674	psi // kgf/cm^2
Density		32 // 512	lbs/ft^3 // kg/m^3
Maximum crushing strength	3,560 // 250	5,901 // 414	psi // kgf/cm^2
Shearing strength		1,334 // 93	psi // kgf/cm^2
Stiffness	1,188 // 83	1,383 // 97	1,000 psi // 1,000 kgf/cm^2
Weight	31 // 496	25 // 400	lbs/ft^3 // kg/m^3
Radial shrinkage		4.1	%
Tangential shrinkage		6	%

HARPEPHYLLUM CAFFRUM

Description

The Kaffir plum or African wild plum is a rounded-shape tree with a single crown growth form. It grows to about 6-15 m (20-50 ft) with irregular branching, and has terminal clusters of spirally-arranged glossy leaves. This evergreen is used as an ornamental garden tree to attract birds and butterflies

The bark is dark in color and deeply fissured that allows other plants to grow up from the ground and under a canopy.

Beschreibung

Die Afrikanische Wildpflaume ist ein Baum mit runder Form und einzelner Krone. Sie erreicht eine Höhe von 6-15 m, die Äste wachsen in unregelmäßigen Abständen und an ihren Enden wachsen in Büscheln die spiralförmig angeordneten, glänzenden Blätter. Dieser immergrüne Baum taucht als Zierpflanze in Gärten auf und zieht Vögel und Schmetterlinge an.

Die Rinde der Afrikanischen Wildpflaume hat eine dunkle Färbung. Ihre tiefen Furchen ermöglichen anderen Pflanzen, unter dem Schutz ihres Blätterdachs vom Boden hoch zu wachsen.

Scientific / Botanical name
Harpephyllum caffrum

Trade / Common name
Kaffir plum

Family name
Anacardiaceae

Regions / Countries of distribution
Australia

Global threat status
NE

Common uses

It's used for furniture and beams but it's not very durable. The edible red fruit is harvested to make wine and jam. The bark is used for treating skin problems, and powdered burnt bark serves to treat sprains and bone fractures.

Allgemeine Verwendung

Zur Herstellung von Balken und Mobiliar, jedoch nicht sehr beständig. Die essbare rote Frucht wird zur Herstellung von Wein und Marmelade geerntet. Die Rinde dient zur Behandlung von Hautproblemen. In verbrannter Form zu Puder verarbeitet behandelt man mit ihr Verstauchungen und Knochenfrakturen.

Machinability

Boring	
Carving	
Mortising	
Moulding	
Polishing	
Turning	

Physical properties

Numerical data	Green	Dry	English / Metric
Radial shrinkage		4	%
Tangential shrinkage		7	%

RHUS TYPHINA

Description

A member of the Anacardiaceae family, this shrub or small tree with deciduous leaves native to eastern North America is the largest of all of the sumac species. It is commonly found in the form of thicket, forming large groups of male or female plants whose height scarcely reaches 12 m (40 ft). It was introduced in Europe in the 17th century and it is still very popular there as an ornamental plant. A pink juice is extracted from its fruit, but not from those of the other sumacs, which are extremely poisonous. Its wood, which is soft, light and brittle, has a fine sapwood with whitish tones, an orange heartwood and a very lustrous texture.

Beschreibung

Der Essigbaum oder Hirschkolbensumach aus dem Osten Nordamerikas ist Mitglied der Familie der Anacardiaceae und höchster Repräsentant alle Pflanzenarten der Gattung Sumach. Üblicherweise findet man ihn als Strauch in großen Gruppen männlicher oder weiblicher Pflanzen, deren Höhe kaum 12 m überschreitet. Im XVII. Jahrhundert wurde er in Europa eingeführt, wo er bis heute eine beliebte Zierpflanze darstellt. Aus seinen Steinfrüchten – jedoch nicht aus denen der anderen Arten der Sumach, hier sind sie extrem giftig – wird eine rosafarbene Limonade gewonnen. Sein Holz ist weich, leicht und brüchig. Das Splintholz ist zart und weißlich, das Kernholz orangefarben und von sehr glänzender Textur.

Scientific / Botanical name
Rhus typhina

Trade / Common name
Staghorn sumac

Family name
Anacardiaceae

Regions / Countries of distribution
Southeastern Canada, northeastern and midwestern United States

Global threat status
NE

Common uses

As it is such a soft wood, it is suitable for uses that require easy handling, such as carvings, candlesticks, turnery or specialty products.

Allgemeine Verwendung

Weil es so weich ist, lässt sich das Holz leicht bearbeiten und kann daher für Schnitzereien, Kerzenhalter, Holzdrehteile oder andere spezielle Produkte verwendet werden.

Machinability

Boring	
Mortising	
Moulding	
Nailing	
Planing	
Turning	

Physical properties

Numerical data	Green	Dry	English / Metric
Weight		44 // 704	lbs/ft^3 // kg/m^3
Radial shrinkage		4.1	%
Tangential shrinkage		7.6	%

SAPINDUS SAPONARIA

Description

Sapindus saponaria is a tall deciduous tree reaching 24 m (80 ft) in height with a rounded crown and scaly, grayish red bark. Leaves, shiny green above and fuzzy below, can be odd-pinnate with a terminal leaflet or even-pinnate lacking a terminal leaflet. Inflorescence occurs in triangular panicles of about 30 cm (12 in). The drupe-like fruit, called soap nut, is orange-brown, slightly translucent and globular. Saponin is an active ingredient in soap nuts that is reported to have medicinal qualities. The wood is light brown, hard, strong, heavy and closed-grained.

Beschreibung

Sapindus saponaria ist ein großer, sommergrüner Baum, der bis zu 24 m Höhe erreicht. Er hat eine kuppelförmige Krone und eine schuppige, gräuliche Rinde. Die Blätter sind auf der Oberseite glänzend grün und auf der Unterseite flaumig. Manche sind unpaarig gefiedert und haben ein Endblättchen, andere sind paarig gefiedert und haben kein Endblättchen. Der Blütenstand erfolgt in dreieckig verzweigten Rispen von ungefähr 30 cm. Die steinfruchtartige Frucht, genannt Seifennuss, ist orange-braun, leicht durchscheinend und rund. Die Seifennuss enthält den Wirkstoff Saponin, der über medizinische Qualitäten verfügen soll. Das Holz ist hellbraun, hart, stark, schwer und dicht.

Scientific / Botanical name
Sapindus saponaria

Trade / Common name
Wingleaf soapberry

Family name
Sapindaceae

Regions / Countries of distribution
Central America, Latin America, North America, Oceania and Southeast Asia

Global threat status
NE

Common uses

The wood of the *Sapindus saponaria* splits easily into thin strips that can be used to make boxes, crates, and baskets. Other than these simple uses, this wood has very little commercial value.

Allgemeine Verwendung

Das Holz des *Sapindus saponaria* lässt sich leicht in dünne Streifen spalten, die zur Herstellung von Schachteln, Kisten und Körben verwendet werden können. Mit Ausnahme dieses einfachen Verwendungszwecks hat das Holz nur einen geringen kommerziellen Wert.

Machinability

Boring	
Mortising	
Moulding	
Planing	

Physical properties

Numerical data	Green	Dry	English / Metric
Density		51 // 816	lbs/ft^3 // kg/m^3
Weight	50 // 800	40 // 640	lbs/ft^3 // kg/m^3

TOONA CILIATA

Description

Toona ciliata is one of the few deciduous trees native to Australia. It is fast-growing with a spreading habit attaining heights between 20 and 30 m (65.6-98 ft) and a trunk diameter reaching 3 m (10 ft). The bark is dark gray developing shallow reticulate fissures as the specimen ages. Leaves are odd-pinnate, sometimes even-pinnate, each with leaflets that are opposite or alternate, ovate-lanceolate, and pubescent. Cream-colored flowers bloom in terminal panicles. The fruit occurs in the form of oblong, green capsules that turn brown and split into star shape to release winged seeds. The wood is light and fine-grained with a deep reddish color. The sapwood is light pink and distinctive from the heartwood.

Beschreibung

Toona ciliata ist einer der wenigen sommergrünen Bäume, die aus Australien stammen. Er ist schnellwachsend mit ausgebreitetem Wuchs und kann eine Höhe von 20-30 m erreichen. Sein Stammdurchmesser beträgt bis zu 3 m. Die dunkelgraue Rinde entwickelt mit zunehmendem Alter leichte, netzartige Furchen. Die Blätter sind teils unpaarig-, teils paarig gefiedert, mit gegenüberliegenden oder wechselständigen Blättchen, eiförmig-lanzettlich und flaumig. Cremefarbene Blüten blühen in endständigen Rispen. Die Frucht hat eine längliche Kapselform und wechselt von Grün zu Braun, bevor sie sich in eine Sternform aufspaltet und geflügelten Samen freigibt. Das Holz ist leicht mit feiner Maserung und einer dunklen, rötlichen Färbung. Das Splintholz ist hellpink und klar vom Kernholz abgegrenzt.

Scientific / Botanical name
Toona ciliata

Trade / Common name
Australian red cedar

Family name
Meliaceae

Regions / Countries of distribution
Afganistan, Papua New Guinea, and Australia

Global threat status
LC

Common uses

Toona ciliata is known as Australian red cedar and its timber is extensively used for interior decorative applications.

Allgemeine Verwendung

Toona ciliata ist unter dem Namen Australische Rotzeder bekannt und ihr Schnittholz findet intensive Nutzung im dekorativen Innenausbau.

Machinability

Boring	
Gluing	
Mortising	
Moulding	
Nailing	
Painting	
Polishing	
Staining	

Physical properties

Numerical data	Green	Dry	English / Metric
Weight		42 // 672	lbs/ft^3 // kg/m^3
Radial shrinkage		4.1	%
Tangential shrinkage		7.6	%

PRUNUS SEROTINA

Description

Belonging to the Rosaceae family, the black cherry or capulí is a tree native to North America introduced in Europe in the 17th century as an ornamental plant, competing with indigenous species and often displacing them. With heights of about 30 m (100 ft) and diameters of approximately 60 cm (24 in) in the best possible condition, it is the largest of all cherry trees on the American continent. In terms of hardness and resistance, its wood is comparable to that of the yellow birch (*Betuna alleghaniensis*). It is slightly more expensive than the wood of the oak, and highly valued given its quality, character and malleability, accounting for its wide range of uses.

Beschreibung

Die Spätblühende Traubenkirsche, auch Späte Traubenkirsche oder Amerikanische Traubenkirsche genannt, ist Mitglied der Familie der Rosaceae und in Nordamerika heimisch. Im XVII. Jahrhundert wurde sie als Zierpflanze in Europa verbreitet, wo sie mit den heimischen Arten konkurriert und sie teilweise verdrängt. Unter optimalen Bedingungen erreicht sie eine Höhe von um die 30 m und einem Stammdurchmesser von etwa 60 cm, was sie zum größten Kirschbaum auf dem amerikanischen Kontinent macht. Was Härte und Widerstandsfähigkeit angeht, ist Ihr Holz vergleichbar mit dem der Gelb-Birke (*Betuna alleghaniensis*). Es ist leicht höher im Preis als Eichenholz und wird aufgrund seiner Wärme, Persönlichkeit und Formbarkeit sehr geschätzt, alles Eigenschaften, die seine vielseitigen Einsatzmöglichkeiten begründen.

Scientific / Botanical name
Prunus serotina

Trade / Common name
Black cherry

Family name
Rosaceae

Regions / Countries of distribution
Eastern North America

Global threat status
NE

Common uses

Cabinetry, carvings, decorative veneers, furniture, trinkets, turnery, musical instruments, interior paneling, construction materials, moldings, flooring, etc.

Allgemeine Verwendung

Tischlerarbeiten, Schnitzereien, Verkleidungen, Dekoration, Mobiliar, Modeschmuck, Holzdrehteile, Musikinstrumente, Polstermöbel, Konstruktionsmaterial, Leisten, Böden, etc.

Machinability

Boring	
Gluing	
Moulding	
Nailing	
Planing	
Polishing	
Screwing	
Staining	
Steam bending	
Turning	

Physical properties

Numerical data	Green	Dry	English / Metric
Bending strength	7,900 // 555	13,250 // 931	psi // kgf/cm²
Hardness		660 // 299	lbs // kg
Impact strength	38 // 96	36 // 91	in // cm
Maximum crushing strength	3,435 // 241	7,865 // 552	psi // kgf/cm²
Shearing strength		1,700 // 119	psi // kgf/cm²
Stiffness	1,380 // 97	1,655 // 116	1,000 psi // 1,000 kgf/cm²
Weight	46 // 736	36 // 576	lbs/ft³ // kg/m³
Radial shrinkage		4	%
Tangential shrinkage		7	%

CELTIS OCCIDENTALIS

Description

This is a slow-growing tree that produces a great many branches, it reaches heights of between 9 and 15 m (30-50 ft) and diameters ranging from 50 to 90 cm (18-36 in). The sapwood is thick with yellow to green tones and spotted with bluish sap. The heartwood is a grayish yellow or a light brown, almost cinnamon color. The texture of the wood is rough and opaque, with a generally straight grain, though it is occasionally full of twists and turns. The hackberry wood, which is relatively scarce, rather fragile and rot-prone, is not suitable as a construction material. It is, however, found in markets, in the form of firewood or as a surface in the place of ash or elm, which are more expensive.

Beschreibung

Es handelt sich um einen langsamwachsenden Baum mit sehr verzweigter Krone, der eine Höhe zwischen 9 und 15 m erreicht. Der Stammdurchmesser schwankt zwischen 50 und 90 cm. Der breite Splint ist grüngelb gesprenkelt mit Flecken aus bläulichem Saft. Das Kernholz ist entweder gelbgrau oder hellbraun, fast zimtfarben gehalten. Die Textur des Holzes ist rau und opak. Die Maserung ist in der Regel geradlinig, teilweise jedoch verdreht. Das Holz des Zürgelbaumes ist recht knapp, spröde und nicht besonders langlebig. Für die Herstellung von Konstruktionsmaterial ist es ungeeignet. Ungeachtet dessen ist es auf dem Markt in Form von Brennholz oder, alternativ zur teureren Esche oder Ulme, als Verkleidung erhältlich.

Scientific / Botanical name
Celtis occidentalis

Trade / Common name
Common hackberry

Family name
Cannabaceae

Regions / Countries of distribution
North America

Global threat status
NE

Common uses

It is used for packaging, cabinetry, plywood, veneers, planks, panels, barrels and furniture. It also serves as a landscaping tree given its ability to adapt to urban settings.

Allgemeine Verwendung

Man verwendet es für Verpackungen, Tischlerarbeiten, Sperrholz, Verkleidungen, Arbeitsplatten, Paneele, Fässer und Mobiliar. Weil der Zürgelbaum sich gut auf ein städtisches Umfeld einlassen kann, wird er im Landschaftsbau eingesetzt.

Machinability

Boring	
Carving	
Gluing	
Mortising	
Moulding	
Nailing	
Painting	
Planing	
Polishing	
Staining	
Turning	
Varnishing	

Physical properties

Numerical data	Green	Dry	English / Metric
Bending strength	6,500 // 456	11,000 // 773	psi // kgf/cm^2
Hardness		880 // 399	lbs // kg
Maximum crushing strength	2,650 // 186	5,440 // 382	psi // kgf/cm^2
Shearing strength		1,590 // 111	psi // kgf/cm^2
Stiffness	950 // 66	1,190 // 83	1,000 psi // 1,000 kgf/cm^2
Weight	50 // 800	37 // 592	lbs/ft^3 // kg/m^3
Radial shrinkage		5	%
Tangential shrinkage		9	%

ZELKOVA SERRATA

Description

The Japanese zelkova (*Zelkova serrata*) is a medium to large tree generally growing to a height between 15 and 24 m (50-80 ft). The bark, smooth and gray when young, flakes in small patches with age to expose orange-brown inner bark. Its spreading upward branching and deciduous foliage displaying attractive shades of ochre and red in the fall make it a very graceful tree. Leaves are alternate, oblong-elliptic with toothed margins. Small green flowers give way to inconspicuous ovate drupes. The wood has a unique tan color with pattern variations.

Beschreibung

Die Japanische Zelkove (*Zelkova serrata*) ist ein mittelgroßer bis großer Baum, der zwischen 15 und 24 m hoch wachsen kann. Die Rinde eines jungen Baums ist glatt und grau. Mit zunehmendem Alter bröckelt sie in kleinen Stücken ab, sodass darunter die orange-braune Innenrinde zum Vorschein kommt. Die nach oben ausgerichteten Zweige und das sommergrüne Blattwerk lassen den Baum im Herbst in attraktiven Schattierungen von Ocker und Rot sehr anmutig erscheinen. Die Blätter sind wechselständig, länglich-elliptisch mit gezahnten Rändern. Aus kleinen, grünen Blüten entwickeln sich unauffällige, eiförmige Steinfrüchte. Das Holz hat eine einzigartige, bräunliche Farbe mit verschiedenen Mustern.

Scientific / Botanical name
Zelkova serrata

Trade / Common name
Japanese zelkova

Family name
Ulmaceae

Regions / Countries of distribution
Japan, Korea, eastern China, and Taiwan

Global threat status
NE

Common uses

Zelkova serrata is a prized commercial timber in Japan. Its close-grained, high-quality wood is extensively used for the fabrication of fine furniture and tool handles.

Allgemeine Verwendung

Zelkova serrata ist in Japan ein wertvolles Schnittholz. Sein feinfaseriges, hochqualitatives Holz wird intensiv für die Herstellung von Mobiliar und Werkzeuggriffen genutzt.

Machinability

Gluing	
Nailing	
Planing	
Polishing	
Sanding	
Screwing	
Steam bending	

Physical properties

Numerical data	Green	Dry	English / Metric
Bending strength		13,255 // 931	psi // kg/cm^2
Density		39 // 624	lbs/ft^3 // kg/m^3
Hardness		1,065 // 483	lbs // kg
Maximum crushing strength		7,555 // 531	psi // kg/cm^2
Shearing strength		1,505 // 105	psi // kg/cm^2
Stiffness		1,705 // 119	1,000 psi // 1,000 kg/cm^2
Weight		39 // 624	lbs/ft^3 // kg/m^3

PRUNUS AVIUM

Description

The wild cherry is an Eurasian species from which the majority of commercial crops are obtained, concentrated in many of the world's temperate regions. Its height can vary from the size of a shrub —it is often used as a hedge— to medium sizes —18-24 m (60-80 ft)— which is generally the starting point for its use as a timber-yielding species. It is typically available on the European market in the form of moderately priced surfaces. It produces a wood with warm colors —reddish brown— a fine, uniform texture and straight grain, with physical properties that are somewhat inferior to those of beech and comparable to oak.

Beschreibung

Die Vogel-Kirsche ist eine eurasische Spezies, von der sich die meisten existenten kommerziellen Zuchtformen der Kirsche ableiten, die zum Großteil in den gemäßigten Regionen der Welt vorkommen. Die Größe der Vogel-Kirsche variiert zwischen strauchartigen Dimensionen, die oftmals als Hecke eingesetzt werden, bis hin zu mittleren Größen von 18-24 m, von denen Teile zur Nutzholzgewinnung eingesetzt werden. In der Regel findet man ihr Holz auf dem europäischen Markt in Form von Verkleidungen zu moderaten Preisen. Das Holz besitzt warme, rotbraune Farben, ist von zarter, ebenmäßiger Textur und hat eine geradlinige Maserung. Ihre physischen Eigenschaften liegen etwas unterhalb denen der Rotbuche und sind vergleichbar mit denen der Eiche.

Scientific / Botanical name
Prunus avium

Trade / Common name
Wild cherry

Family name
Rosaceae

Regions / Countries of distribution
Europe, northwestern Africa, and western Asia

Global threat status
NE

Common uses

Its wood tends to curl but is highly malleable, which favors its use for veneers and plywood, cabinetry, toys, turnery, carpentry and musical instruments.

Allgemeine Verwendung

Ihr Holz tendiert dazu, sich zu verziehen, ist jedoch sehr formbar, weshalb es vorwiegend für die Herstellung von Verkleidungen, Sperrholz, Tischlerarbeiten, Spielwaren, Holzdrehteilen, Schreinerarbeiten oder Musikinstrumenten eingesetzt wird.

Machinability

Gluing	
Nailing	
Planing	
Polishing	
Staining	
Steam bending	
Turning	

Physical properties

Numerical data	Green	Dry	English / Metric
Bending strength	8,835 // 621	15,105 // 1061	psi // kgf/cm²
Hardness		1,300 // 589	lbs // kg
Impact strength	41 // 104	43 // 109	in // cm
Maximum crushing strength	3,870 // 272	7,594 // 533	psi // kgf/cm²
Shearing strength		2,100 // 147	psi // kgf/cm²
Stiffness	1,284 // 90	1,584 // 111	1,000 psi // 1,000 kgf/cm²
Weight		38 // 608	lbs/ft³ // kg/m³

BROSIMUM ALICASTRUM

Description

This plant species, which shares its family with the ficuses and the mulberries, is native to Mesoamerica. It is known by more than fifty indigenous names, with the most common English names being the breadnut and the Maya nut. It reaches heights of between 35 and 45 m (115-148 ft) and diameters of about a meter (39 in). Its grooved bark produces a sweet, sticky sap. Its wood is exceptionally hard and dense, with properties superior to those of the oak, teak and maple. The texture of its wood is slightly porous, opaque and marked by crooked veins. The heartwood and sapwood, which are indistinguishable from each other, feature yellow to grayish tones.

Beschreibung

Diese botanische Spezies gehört zur selben Familie wie der Ficus und der Maulbeerbaum, er ist endemisch und stammt aus Mesoamerika. Es existieren eine Vielzahl einheimischer Namen, von denen im Deutschen Brotnussbaum und Ramonbaum überwiegen. Er wird zwischen 35 und 45 m hoch und erreicht einen Stammdurchmesser von etwa einem Meter. Seine gerippte Rinde produziert einen süßen, klebrigen Saft. Sein Holz ist von besonderer Härte und Dichte, mit Eigenschaften, die die der Eiche, des Teaks oder des Ahorns in den Schatten stellen. Die Textur des Holzes ist ein wenig porös, opak und mit gedrehten Adern. Kern- und Splintholz sind nicht differenzierbar und von graugelber Farbe.

Scientific / Botanical name
Brosimum alicastrum

Trade / Common name
Breadnut

Family name
Moraceae

Regions / Countries of distribution
Western Central Mexico, southern Mexico, Guatemala, El Salvador, the Caribbean, and the Amazon

Global threat status
NE

Common uses
Although this is an endangered species in some regions, its trade remains undisturbed. It is used in the construction of buildings, furniture, flooring, lightweight objects, veneers, etc.

Allgemeine Verwendung
Obwohl diese Spezies in einigen Regionen bedroht ist, wird sie weiterhin ganz regulär kommerzialisiert. Es wird im Bereich der Gebäudekonstruktion, zur Herstellung von Mobiliar, Böden, Leichtbaukonstruktionen, Verkleidungen, etc. eingesetzt.

Machinability

| Boring |
| Mortising |
| Moulding |
| Planing |
| Polishing |
| Steam bending |
| Turning |

Physical properties

Numerical data	Green	Dry	English / Metric
Bending strength		16,030 // 1,127	psi // kgf/cm^2
Hardness		1,520 // 689	lbs // kg
Maximum crushing strength		8,870 // 623	psi // kgf/cm^2
Stiffness		1,850 // 130	1,000 psi // 1,000 kgf/cm^2
Weight	77 // 1,233	55 // 881	lbs/ft^3 // kg/m^3
Radial shrinkage		5	%
Tangential shrinkage		9	%

PYRUS COMMUNIS

Description

The European pear is a deciduous species of the Rosaceae family whose true range of distribution was initially Eastern Europe and Asia Minor, though today it extends to the temperate regions of all continents, due to the vital economic importance of its fruit. Therefore, its supply is limited to old trees, the majority of which goes to the European market. It is a shrub or small tree species whose dimensions, in the best cases, barely exceed 12 m (40 ft) in height and 60 cm (24 in) in diameter. It produces a wood that is pinkish in color, spotted, sturdy and easily sawed, aesthetically similar to the ebony and physically comparable to the oak.

Beschreibung

Die Kultur-Birne ist eine laubabwerfende Spezies aus der Familie der Rosaceae. Ihre ursprünglichen Verbreitungsgebiete wurden in Osteuropa und Kleinasien entdeckt. Dank der lebenswichtigen wirtschaftlichen Relevanz ihrer Frucht verbreitete sie sich in allen gemäßigten Regionen sämtlicher Kontinente. Aus diesem Grund ist ihr Vertrieb auf alte Bäume beschränkt, die zum Großteil für den europäischen Markt bestimmt sind. Es handelt sich um eine Spezies, die strauchartig oder als kleiner Baum auftritt. Ihre Dimensionen überschreiten im besten Fall kaum die Grenze von 12 m Höhe und etwa 60 cm Stammdurchmesser. Sie produziert ein rosenrotes Holz, gesprenkelt, robust und auswalzbar. Vom ästhetischen Aspekt her ähnlich dem Ebenholz, vom physischen Aspekt her mit der Eiche vergleichbar.

Scientific / Botanical name
Pyrus communis

Trade / Common name
European pear

Family name
Rosaceae

Regions / Countries of distribution
Eastern Europe and southwest Asia

Global threat status
NE

Common uses

Its wood is typically used for applications that require malleability and resistance: etchings, turnery, marquetry, musical instruments, tool handles, decorative veneers, etc.

Allgemeine Verwendung

Ihr Holz wird in der Regel in Bereichen eingesetzt, in denen Formbarkeit und Widerstandsfähigkeit gefragt sind: Prägungen, Holzdrehteile, Intarsie, Musikinstrumente, Werkzeuggriffe, dekorative Verkleidungen, etc.

Machinability

Carving	
Gluing	
Nailing	
Planing	
Polishing	
Sanding	
Screwing	
Staining	
Turning	

Physical properties

Numerical data	Green	Dry	English / Metric
Hardness		1,660 // 752	lbs // kg
Maximum crushing strength		6,400 // 449	psi // kgf/cm^2
Shearing strength		12,080 // 849	psi // kgf/cm^2
Weight		45 // 720	lbs/ft^3 // kg/m^3
Radial shrinkage		3.9	%
Tangential shrinkage		11.3	%

CELTIS AFRICANA

Description

This is a tree from the Cannabaceae family whose range extends across large areas of the southern and eastern parts of Southern Africa. It is a polymorphic species given the tremendous variability of the ecosystems in which it thrives: it ranges from shrub sizes in arid and rocky regions, to as tall as 12 m (3 ft) in the savannah and up to 25 m (82 ft) in the areas with the greatest forest concentration. It is known as the white stinkwood (due to the foul smell of its freshly cut wood), which has led it to be confused with the Cape laurel (*Ocotea bullata* or stinkwood), a species with high commercial value with which it has nothing in common, except for the unpleasant odor.

Beschreibung

Baum aus der Familie der Cannabaceae, der in weiten Teilen des südlichen und östlichen Südafrikas verbreitet ist. Aufgrund der Vielzahl an unterschiedlichen Ökosystemen in denen sie wächst, wird diese Spezies als polymorph bezeichnet: sie schwankt zwischen strauchartigen Größen in trockenen, felsigen Gegenden, über Größen von 12 m in Savannen, bis hin zu Größen von 25 m in bewaldeten Gebieten. Im Englischen ist die Spezies unter dem Namen *white stinkwood* bekannt (ihr frisch geschlagenes Holz ist übelriechend), was häufig dazu führt, dass sie mit dem Stinkwood (*Ocotea bullata*) verwechselt wird, einer Spezies mit hohem Marktwert, mit der sie, bis auf den unliebsamen Ausfluss, nichts gemeinsam hat.

Scientific / Botanical name
Celtis africana

Trade / Common name
White stinkwood

Family name
Cannabaceae

Regions / Countries of distribution
South and southeastern Africa

Global threat status
NE

Common uses

Its wood is not commercially relevant because, despite the fact that it is hard and resistant, it is extremely difficult to work with. However, it is possible to find it in the form of shelves, furniture or planks.

Allgemeine Verwendung

Weil es trotz seiner Härte und Widerstandsfähigkeit sehr schwer zu bearbeiten ist, hat das Holz des White Stinkwood kommerziell keine Relevanz. Trotz alledem kann man es unter Umständen in Form von Regalen, Mobiliar oder Arbeitsplatten antreffen.

Machinability

Boring	
Gluing	
Nailing	
Planing	
Polishing	
Sanding	
Screwing	
Staining	
Steam bending	
Turning	
Varnishing	

Physical properties

Numerical data	Green	Dry	English / Metric
Bending strength	12,662 // 890	19,586 // 1,377	psi // kgf/cm²
Density		44 // 704	lbs/ft³ // kg/m³
Hardness		1,986 // 900	lbs // kg
Impact strength		54 // 137	in // cm
Maximum crushing strength	5,408 // 380	8,171 // 574	psi // kgf/cm²
Shearing strength		2,852 // 200	psi // kgf/cm²
Stiffness	1,473 // 103	1,701 // 119	1,000 psi // 1,000 kgf/cm²
Weight	43 // 688	35 // 560	lbs/ft³ // kg/m³
Radial shrinkage		4.1	%
Tangential shrinkage		6	%

MACLURA POMIFERA

Description

A member of the Moraceae family, the Osage orange or hedge-apple is a species native to southern North America though it has spread to all of the United States and Ontario, Canada, as a result of its adaptability. It is a small tree that rarely exceeds 15 m (50 ft) in height and 60 cm (22 in) in diameter. It produces a dense, heavy, knotty wood, with very tight grain and tones ranging from yellowish to orange, suitable for all types of applications that require stability, resistance and durability —it is considered the most rot-proof species in the United States. Although it is possible to use its wood for small projects, its importance as a timber-yielding species is marginal.

Beschreibung

Der Milchorangenbaum, auch Osagedorn genannt, gehört zur Familie der Moraceae und stammt ursprünglich aus dem Süden Nordamerikas. Aufgrund seiner Anpassungsfähigkeit ist er im gesamten Gebiet der Vereinigten Staaten und in Ontario (Kanada) verbreitet. Es handelt sich um einen kleinen Baum, der selten die Höhe von 15 m und den Stammdurchmesser von 60 cm überschreitet. Er produziert ein dichtes, schweres und astiges Holz mit einer sehr engen Maserung in Gelb- und Orangetönen. Es eignet sich für Zwecke, die ein stabiles, widerstandsfähiges und langlebiges Holz erfordern. Es gilt als die unverweslichste Spezies in den Vereinigten Staaten. Obwohl die Möglichkeit besteht, für kleinere Projekte an das Holz zu gelangen, ist diese Spezies für die Nutzholzgewinnung unbedeutend.

Scientific / Botanical name
Maclura pomifera

Trade / Common name
Hedge-apple

Family name
Moraceae

Regions / Countries of distribution
Southeastern North America

Global threat status
NE

Common uses

Its exceptional resistance to humidity makes it the best option for the construction of posts, fences and stakes. It also works perfectly as a hedge or ornamental tree.

Allgemeine Verwendung

Seine außergewöhnliche Resistenz gegenüber Feuchtigkeit macht es zur besten Option bei der Herstellung von Pfosten, Zäunen oder Pflöcken. Zudem dient der Milchorangenbaum perfekt als Hecke oder Ziergehölz.

Machinability

Gluing	
Nailing	
Planing	
Staining	

Physical properties

Numerical data	Green	Dry	English / Metric
Hardness	2,040 // 925	2,760 // 1251	lbs // kg
Maximum crushing strength		9,380 // 659	psi // kgf/cm^2
Weight	62 // 993	56 // 897	lbs/ft^3 // kg/m^3
Volumetric shrinkage		9	%

ARTOCARPUS ALTILIS

Description

Known as the breadfruit, this rapidly growing evergreen species plays a crucial dietary and agricultural role not only in Southeast Asia —where it originated— but also in practically the entire Pacific tropical region —where it spread in the 18th century. It is a species with great variability (there are hundreds of known varieties) and adaptability. It grows to heights of 15-21 m (48-70 ft), with trunks measuring up to 2 m (6,6 ft) in diameter. A milky latex covers the full length of its trunk. Although it is a light, soft, low-density wood, it is quite firm and resistant. It has a porous texture with no visible growth rings. Its poor physical properties inhibit its extensive commercial use.

Beschreibung

Bekannt als Brotfruchtbaum ist diese schnellwachsende, immergrüne Spezies nicht ausschließlich wichtiger Bestandteil der Ernährung und der Landwirtschaft seiner Heimat Südostasien sondern praktisch auch jedes pantropischen Gebiets am Pazifik, wo er seit dem XVIII.Jahrhundert verbreitet wurde. Es handelt sich um eine Spezies von großer Variabilität (es sind hundert Sorten bekannt) und Anpassungsfähigkeit. Er wächst bis zu einer Höhe von 15-21 m, wobei sein Stammdurchmesser bis zu 2 m betragen kann. Milchiges Latex bedeckt seinen Stamm. Obwohl das Holz leicht, weich und von geringer Dichte ist, ist es beachtlich stark und widerstandsfähig. Es hat eine poröse Textur und keine sichtbaren Wachstumsringe. Seine schlechten physischen Eigenschaften verbieten eine extensive kommerzielle Nutzung.

Scientific / Botanical name
Artocarpus altilis

Trade / Common name
Breadfruit

Family name
Moraceae

Regions / Countries of distribution
Southeast Asia, and Pacific Ocean Islands

Global threat status
NE

Common uses

While it is not a widely commercialized wood, it can be found in surfboards, crates, canoes and crafts, and it suitable for lightweight construction, particleboard and wood pulp.

Allgemeine Verwendung

Wenngleich es auf dem Markt nur einen kleinen Anteil einnimmt, ist es in Form von Surfbrettern, Kisten, Kanus oder im Kunstgewerbe zu finden. Zudem kann es für die Herstellung von Leichtbaukonstruktionen, Rohspanplatten oder Zellstoff verwendet werden.

Machinability

Boring	
Mortising	
Moulding	
Planing	
Polishing	
Sanding	
Turning	

Physical properties

Numerical data	Green	Dry	English / Metric
Bending strength	6,111 // 429	9,589 // 674	psi // kgf/cm^2
Maximum crushing strength	2,139 // 150	3,916 // 275	psi // kgf/cm^2
Weight	21 // 336	17 // 272	lbs/ft^3 // kg/m^3
Radial shrinkage		2	%
Tangential shrinkage		5	%

CELTIS AUSTRALIS

Description

The European nettle tree or Mediterranean hackberry is a deciduous tree with a straight shaft and ashen bark that is free of grooves or cracks. Native to the Mediterranean basin but naturalized in nearly all of the world's temperate zones, it is found in clearings along cliffs, rivers or streams, or isolated in chalky or rocky soil. In ideal conditions, it reaches heights of between 20 and 25 m (66-82 ft) and diameters of up to 4 m (13 ft). It produces a wood with straight grain and a fine, opaque texture, with the sapwood and heartwood subsumed in a background of sky blue tones. It is a rather light but resistant and malleable material, which makes it suitable for the production of all types of tools.

Beschreibung

Der Europäische Zürgelbaum oder Südliche Zürgelbaum ist ein laubabwerfender Baum mit geradem Stamm und aschgrauer Rinde, frei von Rillen und Spalten. Ursprünglich stammt er aus dem Mittelmeerraum, wurde jedoch in fast allen gemäßigten Gebieten der Erde heimisch gemacht. Er wächst in Waldgebieten, nahe von Felsschluchten oder an Flussufern und Bächen oder allein auf kalkhaltigen oder steinigen Böden. Unter optimalen Bedingungen erreicht er eine Höhe zwischen 20 und 25 m und einen Stammdurchmesser von bis zu 4 m. Sein Holz hat eine geradlinige Maserung, eine zarte Textur und ist opak. Splint- und Kernholz ordnen sich einem azurblauen Grundton unter. Es handelt sich um ein recht leichtes und dennoch widerstandsfähiges und formbares Material, was es ideal für die Herstellung jeder Art von Werkzeug macht.

Scientific / Botanical name
Celtis australis

Trade / Common name
European nettle tree

Family name
Ulmaceae

Regions / Countries of distribution
Mediterranean region

Global threat status
NE

Common uses

The excellent relationship between resistance and weight makes this wood a widely used material in turnery and carpentry. A yellow dye used for silk is obtained from its roots and bark.

Allgemeine Verwendung

Die exzellente Ausgewogenheit von Widerstandsfähigkeit und Gewicht haben das Holz zu einem häufig eingesetzten Material für Holzdrehteile und Schreinerarbeiten gemacht. Aus Wurzeln und Rinde wird eine gelbe Tinktur zur Einfärbung von Seide gewonnen.

Machinability

Boring	
Mortising	
Moulding	
Planing	
Turning	

Physical properties

Numerical data	Green	Dry	English / Metric
Density		40 // 640	lbs/ft^3 // kg/m^3
Weight	39 // 624	31 // 496	lbs/ft^3 // kg/m^3
Radial shrinkage		4.1	%
Tangential shrinkage		6	%

CELTIS SINENSIS

Description

The Chinese hackberry, a member of the Cannabaceae family, is a deciduous species native to East Asia that is very similar to its American counterpart (*Celtis occidentalis*) though smaller in size —with heights ranging from 12 m (40 ft) to 18 m (60 ft). Its natural range of distribution includes China, Japan, Korea and Taiwan, but its appeal as an ornamental tree, enhanced by its rapid growth and easy cultivation, has led it to spread to areas such as the United States and Australia, where it is considered an invasive species. The wood that it produces is naturally resistant but has poor physical properties: extremely fragile and weak, it has no commercial viability.

Beschreibung

Der Chinesische Zürgelbaum gehört zur Familie der Cannabaceae. Er ist eine laubabwerfende Spezies, die im Osten Asiens beheimatet ist. Trotz seiner geringeren Höhe, zwischen 12 m und 18 m, ist er seinem amerikanischen Pendant (*Celtis occidentales*) sehr ähnlich. Sein natürliches Verbreitungsgebiet erstreckt sich über China, Japan, Korea und Taiwan. Die Gebiete in denen er als schnellwachsendes und leicht kultivierbares Ziergehölz beliebt ist, liegen jedoch bis in die Vereinigten Staaten oder Australien verstreut, wo man den Chinesischen Zürgelbaum als invasive Art betrachtet. Er produziert ein von Natur aus widerstandsfähiges Holz mit schlechten physischen Eigenschaften: Extrem spröde und schwach, was ihn auf kommerzieller Ebene unbrauchbar macht.

Scientific / Botanical name
Celtis sinensis

Trade / Common name
Chinese hackberry

Family name
Cannabaceae

Regions / Countries of distribution
East Asia

Global threat status
NE

Common uses

Its cultivation, which is relatively common, is intended to encourage the market for ornamental trees, urban species and household species, in the form of bonsai.

Allgemeine Verwendung

Seine Kultivierung zielt im Allgemeinen auf die Sättigung des Marktes für Zierhölzer ab: für die Gestaltung der Städte, der heimischen Gärten oder in Form eines Bonsais.

Machinability

Boring	
Gluing	
Mortising	
Moulding	
Planing	

Physical properties

Numerical data	Green	Dry	English / Metric
Radial shrinkage		4	%
Tangential shrinkage		7	%

HEMIPTELEA DAVIDII

Description

The abura, also known as thorned elm and simu namu, is a deciduous tree that grows up to 15 m (50 ft). It is related to elms and zelkovas. Generally round-shaped and sometimes spreading horizontally, this tree has elliptic and toothed leaves arranged opposite one another and produces small whitish flowers similar to those of the elm trees.

The grayish brown bark is used to manufacture textiles and the leaves are used as tea for their diuretic qualities. Seed oil is extracted but reports do not confirm if the oil is edible or serves any use.

Beschreibung

Die Dornulme ist ein sommergrüner Baum, der bis zu 15 m hoch wird. Sie ist mit Ulmen und Zelkoven verwandt. In der Regel rund geformt, breitet sich die Dornulme manchmal horizontal aus. Ihre elliptischen, gezahnten Blätter sind gegenüberliegend angeordnet und ihre weißlichen Blüten sind denen der Ulme ähnlich.

Die gräulich braune Rinde wird zur Herstellung von Textilien verwendet und ihre Blätter haben als Tee harntreibende Qualitäten. Aus ihren Samen wird Öl gewonnen, von dem man jedoch nicht mit Gewissheit sagen kann, ob es genießbar ist oder einen anderweitigen Nutzen hat.

Scientific / Botanical name
Hemiptelea davidii

Trade / Common name
Abura

Family name
Ulmaceae

Regions / Countries of distribution
China and Korea

Global threat status
NE

Common uses

The hardwood of the *Hemiptelea davidii* is used for making tools.

Allgemeine Verwendung

Das Hartholz von *Hemiptelea davidii* wird zur Herstellung von Werkzeugen verwendet.

Machinability

Boring	
Mortising	
Moulding	
Planing	
Polishing	
Turning	

Physical properties

Numerical data	Green	Dry	English / Metric
Radial shrinkage		4	%
Tangential shrinkage		8	%

ZIZIPHUS JUJUBA

Description

Ziziphus jujuba, commonly known as jujube, is an attractive deciduous tree or shrub with contorted, thorny branches growing 3 to 6 m (10-20 ft). The bark is gray. The leaves of the jujube are shiny green, alternate, ovate-acute with subtly serrated margins. Flowers, small and yellow, appear in clusters of 2 to 3 along the entire length of the branches. The edible fruit is a globose smooth-skinned drupe, green-colored when immature turning brown when ripe. The jujube is one of the most cultivated trees, especially in China where its fruits are much prized.

Beschreibung

Ziziphus jujuba, gemeinhin als Chinesische Jujube bezeichnet, ist ein attraktiver sommergrüner Baum oder Busch mit gebogenen, dornenreichen Ästen, der 3 bis 6 m hoch wird. Die Rinde ist grau. Die Blätter der Chinesischen Jujube sind leuchtend grün, wechselständig eiförmig-spitz, mit fast unmerklich geriffelten Rändern. Die kleinen, gelben Blüten sind in 2er- und 3er-Formationen zusammengefasst und über die gesamte Astlänge verteilt. Die kugelförmige Steinfrucht hat eine glatte Schale, die zunächst grün ist und im Reifeprozess braun wird. Die Chinesische Jujube ist einer der am häufigsten kultivierten Bäume. Dies gilt vor allem für China, wo seine Früchte sehr geschätzt werden.

Scientific / Botanical name
Ziziphus jujuba

Trade / Common name
Jujube

Family name
Rhamnaceae

Regions / Countries of distribution
Africa, Oceania and Southeast Asia

Global threat status
LC

Common uses

The wood of the jujube is hard, dense and tough. It is used for making musical instruments such as flutes and oboes.

Allgemeine Verwendung

Das Holz der Chinesischen Jujube ist hart, dicht und stark. Es wird zur Herstellung von Musikinstrumenten wie Flöten oder Oboen verwendet.

Machinability

Boring	
Mortising	
Moulding	
Planing	

Physical properties

Numerical data	Green	Dry	English / Metric
Bending strength	13,580 // 954	20,960 // 1,473	psi // kgf/cm^2
Density		40 // 512	lbs/ft^3 // kg/m^3
Maximum crushing strength	3,915 // 275	6,400 // 449	psi // kgf/cm^2
Stiffness	1,020 // 71	1,200 // 84	1,000 psi // 1,000 kgf/cm^2
Weight	40 // 640	32 // 512	lbs/ft^3 // kg/m^3

OCHROMA PYRAMIDALE

Description

Commonly known as the balsa, this large, rapidly growing tree —it grows to heights of between 18 and 27 m (60-90 ft) and diameters of between 75 and 120 cm (30-48 in)— is a first-rate timber-yielding species. Its very large natural range of distribution (South America and Central America), and the near-ubiquity of its cultivation in tropical areas make its wood —soft and light, with straight and open grain, a very lustrous texture and whitish tones, splashed with yellowish hues— a very abundant material despite its very localized supply. Ecuador exports 95% of the balsa wood consumed in the global market.

Beschreibung

Im Allgemeinen ist dieser große, schnellwachsende Baum als Balsabaum bekannt. Er erreicht eine Höhe zwischen 18 und 27 m und einen Stammdurchmesser zwischen 75 und 120 cm und ist eine Spezies zur Nutzholzgewinnung ersten Ranges. Es ist weich, leicht, mit geradliniger, offener Maserung und von glänzend weißlicher Textur mit gelblichen Flecken. Aufgrund der enormen Größe seines natürlichen Verbreitungsgebietes, seiner praktisch allgegenwärtigen Kultivierung in tropischen Gebieten und trotz seiner sehr lokalisierten Vertreibung, ist sein Holz reichlich verfügbar. Ecuador exportiert 95% des auf dem Weltmarkt nachgefragten Balsaholzes.

Scientific / Botanical name
Ochroma pyramidale

Trade / Common name
Balsa tree

Family name
Malvaceae

Regions / Countries of distribution
Southern Brazil and Bolivia, and Mexico

Global threat status
NE

Common uses
It is a very malleable wood, which allows for a wide range of uses: musical instruments, barrels, boats, drumsticks, turnery, planks, etc.

Allgemeine Verwendung
Es handelt sich um ein sehr formbares Holz, das vielseitig Verwendung findet: Musikinstrumente, Fässer, Schiffbau, Trommelstöcke, Holzdrehteile, Arbeitsplatten, etc.

Machinability

- Carving
- Gluing
- Mortising
- Moulding
- Nailing
- Planing
- Routing and recessing
- Sanding
- Screwing
- Turning

Physical properties

Numerical data	Green	Dry	English / Metric
Bending strength	4,466 // 313	7,325 // 514	psi // kgf/cm^2
Density		10 // 160	lbs/ft^3 // kg/m^3
Hardness		87 // 39	lbs // kg
Maximum crushing strength	2,404 // 169	3,399 // 238	psi // kgf/cm^2
Stiffness	668 // 46	930 // 65	1,000 psi // 1,000 kgf/cm^2
Weight		11 // 176	lbs/ft^3 // kg/m^3
Radial shrinkage		3	%
Tangential shrinkage		8	%

CEIBA PENTANDRA

Description

A member of the Malvaceae family, the ceiba, lupuna or kapok (in Anglophone countries) is a polymorphic species native to the tropics that is capable of growing up to 60 m (200 ft) in height and 1.8 m (72 in) in diameter. It is an important source of wood in Central America and South America, from whence it is shipped to China and Japan. It is aesthetically similar to balsa wood, though not as highly valued, as it is considered too soft and fragile to serve as a material for construction or cabinetry. A fiber eight times lighter than cotton, which functions as an extraordinary thermal and acoustic insulator, is extracted from its wood.

Beschreibung

Der Kapokbaum, auch Wollbaum genannt, ist eine polymorphe Spezies und Mitglied der Familie der Malvaceae. Ursprünglich stammt er aus dem tropischen Regenwald und kann bis zu 60 m hoch werden, wobei sein Stammdurchmesser 1,8 m beträgt. In Mittel- und Südamerika stellt er eine wichtige Holzquelle dar. Von hier aus exportiert man nach China und Japan. Aus ästhetischer Sicht ist das Holz des Kapokbaumes dem des Balsabaumes ähnlich, wobei es nicht so hoch gehandelt wird wie dieser. Da es als sehr weich und spröde eingestuft wird, kann es als Konstruktionsmaterial oder für Tischlerarbeiten eingesetzt werden. Aus seinem Holz kann eine Faser hergestellt werden, die acht Mal leichter ist als Baumwolle und die zudem außergewöhnlich wärme- und lärmisolierend ist.

Scientific / Botanical name
Ceiba pentandra

Trade / Common name
Kapok

Family name
Malvaceae

Regions / Countries of distribution
Mexico, Central America, northern South America, and the Caribbean

Global threat status
NE

Common uses

With no particular relevance to the timber industry, its wood is still used in secondary products: panels, moldings, packaging, barrels, benches and stools, etc.

Allgemeine Verwendung

Ohne über spezielle Relevanz für den Holzsektor zu verfügen, wird sein Holz dennoch weiterhin zur Herstellung von Nebenprodukten eingesetzt: Paneele, Formen, Verpackungen, Fässer, Banken und Hocker, etc.

Machinability

Carving	
Gluing	
Mortising	
Moulding	
Nailing	
Painting	
Planing	
Polishing	
Resistance to splitting	
Sanding	
Steam bending	
Turning	
Veneering qualities	

Physical properties

Numerical data	Green	Dry	English / Metric
Bending strength	2,574 // 180	5,594 // 393	psi // kg/cm^2
Density		20 // 320	lbs/ft^3 // kg/m^3
Hardness		238 // 107	lbs // kg
Maximum crushing strength		3,045 // 214	psi // kg/cm^2
Shearing strength		498 // 35	psi // kg/cm^2
Stiffness	384 // 26	595 // 41	1,000 psi // 1,000 kg/cm^2
Weight	56 // 897		lbs/ft^3 // kg/m^3
Radial shrinkage		3	%
Tangential shrinkage		5	%

CHORISIA SPECIOSA

Description

A member of the same family as the baobab and the kapok, the silk floss tree is a deciduous tree native to the tropical and subtropical rainforests of South America. It has a single bulky trunk protected by conical spines —where water is stored during droughts— which can measure 2 m (79 in) in diameter. In exceptional conditions, it can reach heights of up to 25 m (82 ft), but it typically registers between 6 and 12 m (19-39 ft). Its pods produce a cotton that, while not as high-quality as that of the kapok, has some industrial uses. It is a species that can potentially be used for timber; its light, flexible wood, however, is practically irrelevant commercially.

Beschreibung

Der Florettseidenbaum, auch Chorisie genannt, ist ein laubabwerfender Baum. Er ist in den tropischen und subtropischen Regenwäldern Südamerikas beheimatet und Mitglied der selben Familie wie der Affenbrotbaum und der Wollbaum. Er verfügt über einen einzelnen, ausladenden Stamm mit einem Durchmesser von bis zu 2 m, der mit kegelförmigen Stacheln geschützt ist und in dem er in Zeiten der Dürre Wasser speichern kann. Unter optimalen Bedingungen erreicht er eine Höhe von bis zu 25 m, wobei er in der Regel zwischen 6 und 12 m hoch wächst. Seine Kapselfrüchte produzieren eine Baumwolle, die industriell genutzt wird. Der Ertrag ist jedoch geringer als beim Kapokbaum. Es handelt sich um eine Spezies mit Potential zur Nutzholzgewinnung. Sein leichtes und elastisches Holz ist jedoch im kommerziellen Sinne irrelevant.

Scientific / Botanical name
Chorisia speciosa

Trade / Common name
Silk floss tree

Family name
Malvaceae

Regions / Countries of distribution
Tropical and subtropical South America

Global threat status
NE

Common uses

Edible vegetable oils are obtained from its seeds; cotton is obtained from its pods; its wood is used in the production of packaging, wood pulp and paper.

Allgemeine Verwendung

Aus seinen Samen wird Speiseöl gewonnen, seine Kapselfrüchte produzieren Baumwolle und sein Holz wird zur Herstellung von Verpackungen, Zellstoff und Papier verwendet.

Machinability

Boring	
Mortising	
Moulding	
Planing	
Polishing	
Turning	

Physical properties

Numerical data	Green	Dry	English / Metric
Bending strength	3,229 // 227	4,817 // 338	psi // kgf/cm^2
Density		25 // 400	lbs/ft^3 // kg/m^3
Hardness		319 // 144	lbs // kg
Maximum crushing strength	724 // 50	1,435 // 100	psi // kgf/cm^2
Shearing strength		882 // 62	psi // kgf/cm^2
Stiffness	1,017 // 71	1,199 // 84	1,000 psi // 1,000 kgf/cm^2
Weight	25 // 400	21 // 336	lbs/ft^3 // kg/m^3
Radial shrinkage		3	%
Tangential shrinkage		9	%

TILIA AMERICANA

Description

Tilia americana is commonly known as American linden and basswood among other denominations. Basswood is a fast-growing tree reaching heights between 12 and 15 m (40-50 ft) with an oval crown and a tall straight trunk. The bark is silver gray, finely ridged. Leaves are alternately-arranged, heart-shaped, serrated, green above and paler below. The flowers of the basswood are creamy white and borne in 3 to 7 pendulous cymes. They attract large numbers of bees, which produce a distinctive tasting honey. The fruit is a hard pea-sized nut usually containing one seed. The wood has a uniform cream color showing subtle growth rings and an even texture.

Beschreibung

Tilia americana wird gemeinhin als Amerikanische Linde bezeichnet. Die Amerikanische Linde ist ein schnellwachsender Baum, der eine Höhe zwischen 12 und 15 m erreichen kann. Sie hat eine ovale Krone und einen hohen, geraden Stamm. Die Rinde ist silbergrau und fein gezahnt. Ihre Blätter sind wechselständig angeordnet, herzförmig, geriffelt, oben grün und unten blasser. Die Blüten der Amerikanischen Linde sind cremeweiß und zu 3 bis 7 hängenden Zymen angeordnet. Sie ziehen große Mengen von Bienen an, die einen charakteristisch schmeckenden Honig produzieren. Die Frucht ist eine erbsengroße, harte Nuss, die in der Regel Samen enthält. Das Holz ist einheitlich cremefarben mit feinen Wachstumsringen und einer gleichmäßigen Struktur.

Scientific / Botanical name
Tilia americana

Trade / Common name
American linden

Family name
Tiliaceae

Regions / Countries of distribution
Eastern North America

Global threat status
NE

Common uses

Basswood is an important timber tree. Its easy to work, soft light wood is valued for turned and carved objects. The inner bark is a source of fiber for rope, fish nets and baskets.

Allgemeine Verwendung

Das Holz der Amerikanischen Linde ist ein bedeutendes Nutzholz. Das weiche, leichte Holz ist gut zu bearbeiten und ideal für das Drehen und Schnitzen von Objekten. Die Innenrinde ist Faserquelle zur Herstellung von Tauen, Fischernetzen und Körben.

Machinability

Boring	■
Carving	■
Gluing	■
Mortising	■
Moulding	■
Nailing	■
Painting	■
Planing	■
Polishing	■
Staining	■
Turning	■

Physical properties

Numerical data	Green	Dry	English / Metric
Bending strength	5,100 // 358	8,700 // 611	psi // kg/cm^2
Hardness		410 // 185	lbs // kg
Impact strength	19 // 48	22 // 55	in // cm
Maximum crushing strength	2,420 // 170	4,340 // 305	psi // kg/cm^2
Shearing strength		990 // 69	psi // kg/cm^2
Stiffness	1,017 // 71	1,507 // 105	1,000 psi // 1,000 kg/cm^2
Weight	45 // 720	28 // 448	lbs/ft^3 // kg/m^3
Radial shrinkage		7	%
Tangential shrinkage		9	%

TILIA VULGARIS

Description

Tilia vulgaris is commonly called European lime and common linden. Its height ranges between 19 and 40 m (65-131 ft) and its trunk diameter can reach 2 m (6.5 ft). The bark is light gray and soft on mature specimens and becomes brownish gray with perpendicular fissures. The leaves of this deciduous tree are alternately arranged, heart-shaped, dark green above, paler below, and sharply serrated. The yellowish white flowers hang in flattened clusters from long stalks. The fruit is a nut, fuzzy at first, turning hard and smooth when mature. The wood of *Tilia vulgaris* is whitish, smooth and close-grained with faint signs of the growth rings.

Beschreibung

Tilia vulgaris wird gemeinhin als Holländische Linde bezeichnet. Ihre Höhe variiert zwischen 19 und 40 m und ihr Stamm kann bis zu 2 m Durchmesser erreichen. Die Rinde des jungen Baums ist hellgrau und weich, die des älteren nimmt eine bräunlich graue Färbung an und entwickelt senkrechte Furchen. Die Blätter dieses sommergrünen Baums sind wechselständig angeordnet, herzförmig und scharf gezackt. Auf der oberen Seite haben sie eine dunkelgrüne Farbgebung, die auf der Unterseite blasser ist. Die gelblich-weißen Blüten hängen in flachen Büscheln von langen Stielen herab. Die Frucht der Holländischen Linde ist eine Nuss, die zunächst flaumig, mit zunehmender Reife jedoch hart und glatt ist. Das Holz des *Tilia vulgaris* ist weißlich, glatt und hat eine feine Maserung mit schwach angedeuteten Wachstumsringen.

Scientific / Botanical name
Tilia vulgaris

Trade / Common name
Common lime

Family name
Tiliaceae

Regions / Countries of distribution
Europe

Global threat status
NE

Common uses

The European lime has a lightweight wood very suitable for many purposes including turnery, carving, and the manufacture of sounding boards for pianos. The inner bark is strong and elastic, and good for making rope and fishing nets.

Allgemeine Verwendung

Das Drehen und Schnitzen von Objekten und die Herstellung von Resonanzböden für Klaviere sind nur wenige Beispiele für die Vielzahl von Bereichen, in denen das leichte Holz der Holländischen Linde eingesetzt werden kann. Aufgrund ihrer Stärke und Elastizität ist die Innenrinde ideal für die Herstellung von Tauen und Fischernetzen.

Machinability

Boring	
Carving	
Gluing	
Moulding	
Nailing	
Planing	
Polishing	
Staining	
Steam bending	
Turning	

Physical properties

Numerical data	Green	Dry	English / Metric
Bending strength	7,505 // 527	12,635 // 888	psi // kg/cm^2
Density		34 // 544	lbs/ft^3 // kg/m^3
Hardness		710 // 322	lbs // kg
Impact strength	24 // 60	30 // 76	in // cm
Maximum crushing strength	3,629 // 255	6,634 // 466	psi // kg/cm^2
Shearing strength		1,408 // 98	psi // kg/cm^2
Stiffness	1,423 // 100	1,733 // 121	1,000 psi // 1,000 kg/cm^2
Weight	41 // 656	35 // 560	lbs/ft^3 // kg/m^3
Radial shrinkage		5	%
Tangential shrinkage		7.5	%

HERITIERA LITTORALIS

Description

This much-branched tree averages 15 to 25 m (50-82 ft) and usually has a twisted and stunted bole with wavy buttresses extending far out. Its thin and oblong leaves are arranged in a spiral and two-toned: the top is dark green and the underside is silvery white. The dungun is also known as looking-glass mangrove. It produces small bell-shaped flowers in clusters at the tips of the branches, and shiny brown woody fruits.

The bark is pinkish gray or dark brown, smooth and flaky.

Beschreibung

Dieser vielverzweigte Baum erreicht eine durchschnittliche Höhe von 15-25 m. Für gewöhnlich besitzt er einen gewundenen und verkrümmten Stamm mit gewellter, sehr breit auslaufender Basis. Seine dünnen, länglichen Blätter sind spiralförmig angeordnet und zweifarbig: Die obere Seite ist dunkelgrün, die untere silbrig-weiß. Auch Brillenmangrove genannt. Sie produziert kleine, glockenförmige Blüten, die büschelweise an den Astspitzen wachsen. Ihre Früchte sind glänzend und holzig.

Die Rinde ist entweder pink-grau oder dunkelbraun, glatt und schuppig.

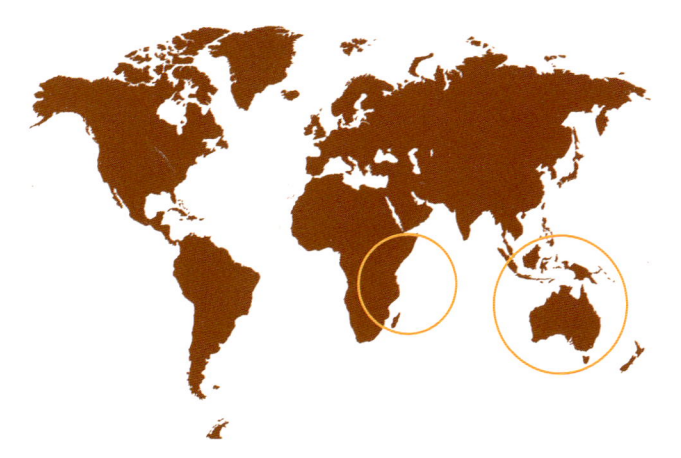

Scientific / Botanical name
Heritiera littoralis

Trade / Common name
Dungun

Family name
Malvaceae

Regions / Countries of distribution
Eastern Africa, Australia and the Pacific

Global threat status
LC

Common uses

The tough wood of the *Heritiera littoralis* has historically been utilized in ship construction, as masts if the trunk is straight enough. It is also used for a wide range of household uses.

Allgemeine Verwendung

Das harte Holz der *Heritiera littoralis* wurde früher bei der Schiffskonstruktion als Mast verwendet, wenn der Stamm gerade genug war. Es wird außerdem häufig für verschiedene Haushaltszwecke verwendet.

Machinability

Boring	
Mortising	
Moulding	
Planing	
Polishing	
Turning	

Physical properties

Numerical data	Green	Dry	English / Metric
Bending strength	11,230 // 789	17,400 // 1,223	psi // kgf/cm^2
Density		51 // 816	lbs/ft^3 // kg/m^3
Hardness		1,660 // 752	lbs // kg
Impact strength		55 // 139	in // cm
Maximum crushing strength	5,905 // 415	8,765 // 616	psi // kgf/cm^2
Shearing strength		2,200 // 154	psi // kgf/cm^2
Stiffness	2,110 // 148	2,440 // 171	1,000 psi // 1,000 kgf/cm^2
Weight	51 // 816	45 // 720	lbs/ft^3 // kg/m^3
Radial shrinkage		4.1	%
Tangential shrinkage		7.6	%

ADANSONIA DIGITATA

Description

Without reaching the height of the sequoias, this African tree with a prehistoric appearance can measure 25 m (82 ft) tall, with trunks up to 40 m (131 ft) in diameter. Commonly known as the baobab, it gets the first half of its scientific name (*Adansonia digitata*) from the botanist Michael Adanson, and the other half from its hand-shaped leaves. It produces a fruit that resembles a small melon. Its trunk can store up to 120 tons of water. A member of the Malvaceae family, it is a deciduous tree with smooth, woody bark and soft, fibrous wood. It has extraordinary longevity: some specimens have lived several thousand years.

Beschreibung

Die Größe der Sequoias ausgenommen, kann dieser afrikanische Baum mit prähistorischem Erscheinungsbild eine Größe von 25 m überschreiten und einen Stammdurchmesser von bis zu 40 m erreichen. Allgemein als Afrikanischer Affenbrotbaum oder Afrikanischer Baobab bekannt, verdankt er den ersten Teil seines wissenschaftlichen Namens (*Adansonia digitata*) dem Botaniker Michael Adanson und den zweiten Teil der Form seiner Blätter, die an eine Hand erinnert. Er produziert Früchte, die wie kleine Melonen aussehen. Sein Stamm kann bis zu 120 Tonnen Wasser speichern. Er gehört zur Familie der Malvaceae und ist ein laubabwerfender Baum mit ebener, holziger Rinde und weichem, faserigem Holz. Er ist außergewöhnlich langlebig: Einige Exemplare sind mehrere Tausend Jahre alt.

Scientific / Botanical name
Adansonia digitata

Trade / Common name
Baobab

Family name
Malvaceae

Regions / Countries of distribution
Africa

Global threat status
NE

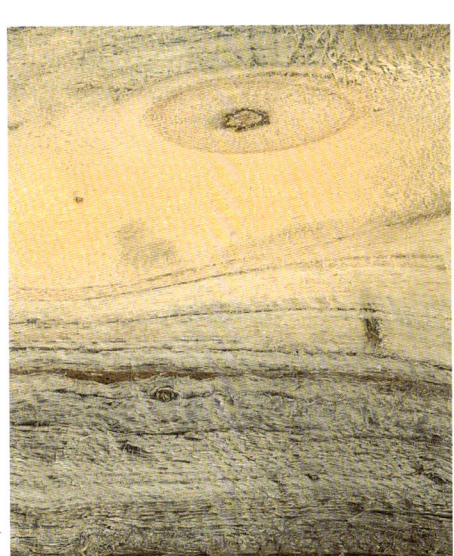

Common uses

Manufacture of paper and rope, oils, enamels, baskets, glue, dyes, fuel, fabrics, strings for musical instruments, boats and canoes, etc.

Allgemeine Verwendung

Zur Herstellung von Papier und Takelage, Öl, Lack, Korb, Klebstoff, Tinte, Brennstoff, Stoff, Saiten für Musikinstrumente, Schiffe und Kanus, etc.

Machinability

Gluing	
Nailing	
Painting	
Planing	
Varnishing	

Physical properties

Numerical data	Green	Dry	English / Metric
Density	17 // 272		lbs/ft^3 // kg/m^3
Weight	17 // 272	14 // 224	lbs/ft^3 // kg/m^3
Radial shrinkage		6	%
Tangential shrinkage		10	%

CAVANILLESIA PLATANIFOLIA

Description

The bongo, macondo, cuipo or hamelí is a large tree —from 45 to 60 m (148-197 ft) in height and between 1.5 and 2 m (59-79 in) in diameter— which thrives in the humid forests of Central America, Colombia and Peru. It is easily identified by both its bulbous trunk, which is smooth, grayish and branchless, and its fruit, a huge edible nut with a peanut flavor. Light and porous, with straight grain, a rough texture and clearly visible growth circles, it is the softest of all known woods. The sapwood and heartwood, only distinguishable when the wood is green, exhibit a subtle transition between brownish-gray tones (sapwood) and ashen tones (heartwood) after oxidation.

Beschreibung

Der Cuipo ist ein sehr großer Baum, der in den Regenwäldern Mittelamerikas, Kolumbiens und Perus wächst. Seine Höhe liegt bei 45-60 m, sein Stammdurchmesser beträgt zwischen 1,5 und 2 m. Er ist leicht anhand seines knolligen, ebenmäßigen Stammes zu erkennen, der gräulich und astfrei ist. Wegen seiner Frucht, einer großen, essbaren Nuss mit Erdnussgeschmack, betet man ihn an. Sein Holz ist leicht und porös, mit geradliniger Maserung, grober Textur und deutlich erkennbaren Wachstumsringen. Von allen bekannten Holzsorten ist es die weichste. Wenn das Holz grün ist, lassen sich Splint- und Kernholz kaum unterscheiden. Sobald eine Oxidation stattgefunden hat, zeigen sie einen raffinierten Übergang von Brauntönen (Splintholz) ins Aschgrau Kernholz.

Scientific / Botanical name
Cavanillesia platanifolia

Trade / Common name
Cuipo tree

Family name
Malvaceae

Regions / Countries of distribution
Tropical Central America

Global threat status
NT

Common uses

Its fibers are used to produce rope; its wood, heavily used locally in the past (canoes and other small boats, cups, tools), is now commercially irrelevant.

Allgemeine Verwendung

Aus seinen Fasern entstehen Seile und Verankerungen, sein Holz war früher auf lokaler Ebene sehr gefragt (Kanus, Boote, Tassen, Werkzeug), wobei es heutzutage im kommerziellen Sinne irrelevant ist.

Machinability

Boring	
Mortising	
Planing	
Sanding	
Turning	

Physical properties

Numerical data	Green	Dry	English / Metric
Density		7 // 112	lbs/ft^3 // kg/m^3
Weight	7 // 112	6 // 96	lbs/ft^3 // kg/m^3
Radial shrinkage		3.3	%
Tangential shrinkage		5.2	%

HIBISCUS TILIACEUS

Description

While *Hibiscus tilaceus* is an evergreen native to the shores of the Pacific and Indian oceans, it is widely cultivated throughout the tropical and subtropical regions as an ornamental tree. It can reach a height of 4 to 10 m (13 to 33 ft) and a trunk diameter up to 15 cm (5.9 in) with a tangled, sprawling form. The sapwood is very narrow and whitish. The gray-brown heartwood is often mottled with gray-blue and purple shades. The bark is gray to light brown, smooth and longitudinally fissured with age.

Beschreibung

Hibiscus tilaceus ist ein immergrüner Baum, der ursprünglich an den Küsten des Pazifiks und des Indischen Ozeans beheimatet ist. Er wird in weiten Teilen der tropischen und subtropischen Regionen als Zierpflanze kultiviert. Er kann eine Höhe von 4-10 m erreichen. Sein Stamm hat einen Durchmesser von bis zu 15 cm und ist verdreht und langgezogen. Sein Splintholz ist schmal und weißlich. Das grau-braune Kernholz ist häufig grau-blau gesprenkelt und hat eine lila Tönung. Die Rinde ist grau bis hellbraun, glatt und mit zunehmendem Alter entwickelt sie senkrechte Furchen.

Scientific / Botanical name
Hibiscus tiliaceus

Trade / Common name
Sea hibiscus

Family name
Malvaceae

Regions / Countries of distribution
Eastern and northern Australia, Oceania, Maldives, and southeast Asia

Global threat status
NE

Common uses

The wood, the bark and the flowers of the sea hibiscus are used for various applications including cabinetmaking, furniture, building construction, firewood and wood carvings. The tough bark is used to make rope and for sealing cracks in boats.

Allgemeine Verwendung

Das Holz, die Rinde und die Blüten des Linden-Roseneibisch werden in verschiedenen Bereichen verwendet: Für die Herstellung von Schränken und anderem Mobiliar, in der Baukonstruktion, als Feuerholz, für Holzschnitzereien. Die harte Rinde wird zur Herstellung von Tauen und zum Flicken von Booten benutzt.

Machinability

Boring	
Mortising	
Moulding	
Planing	
Turning	

Physical properties

Numerical data	Green	Dry	English / Metric
Density		47 // 752	lbs/ft^3 // kg/m^3
Weight	46 // 736	37 // 592	lbs/ft^3 // kg/m^3

THESPESIA POPULNEA

Description

Thespesia populnea is a small evergreen tree typically 6 to 10 m (20-33 ft) in height at maturity with a fairly short, often crooked trunk and a spreading crown. The bark is gray and smooth in younger trees; dark gray and deeply furrowed in older specimens. Leaves are alternate, heart-shaped and shiny dark green. The bell-shaped flowers, similar to those of the hibiscus, are typically yellow with a burgundy base. The fruits of the Portia tree are woody seed capsules that grow on short stalks at the end of branches. The heartwood is reddish to dark brown and the sapwood is cream-colored. The wood texture is fine with straight grain and wavy figure.

Beschreibung

Thespesia populnea ist ein kleiner, immergrüner Baum, der im ausgewachsenen Zustand typischerweise zwischen 6 und 10 m hoch wird. Sein Stamm ist recht kurz und oftmals gekrümmt und seine Krone ausladend. Am jungen Baum ist die Rinde grau und glatt, am älteren Exemplar dunkelgrau und dicht behaart. Die Blätter sind wechselständig, herzförmig und dunkelgrün-glänzend. Die glockenförmigen Blüten sehen denen des Hibiscus ähnlich. Sie sind typischerweise gelb mit einer burgunderroten Basis. Die Früchte des Portiabaums sind holzige Samenkapseln, die an kurzen Stielen am Astende wachsen. Das Kernholz ist rötlich bis dunkelbraun und das Splintholz ist cremefarben. Das Holz verfügt über eine feine Struktur mit gerader Maserung und welligem Muster.

Scientific / Botanical name
Thespesia populnea

Trade / Common name
Portia tree

Family name
Malvaceae

Regions / Countries of distribution
Oceania and Southeast Asia

Global threat status
NE

Common uses

The rich dark wood that the Portia tree produces is carved into small canoes, tools and small artwork, while bark is made into rope.

Allgemeine Verwendung

Das intensiv dunkle Holz des Portiabaums wird zum Bau kleiner Kanus, zur Herstellung von Werkzeug und zur Erstellung kleiner Kunstwerke genutzt. Die Rinde dient zur Erzeugung von Tauen.

Machinability

Boring	
Gluing	
Mortising	
Moulding	
Planing	
Polishing	
Turning	

Physical properties

Numerical data	Green	Dry	English / Metric
Bending strength	9,315 // 654	14,570 // 1,024	psi // kgf/cm^2
Density		45 // 720	lbs/ft^3 // kg/m^3
Impact strength		65 // 165	in // cm
Maximum crushing strength	5,405 // 380	8,175 // 574	psi // kgf/cm^2
Stiffness	1,360 // 95	1,570 // 110	1,000 psi // 1,000 kgf/cm^2
Weight	42 // 672	32 // 512	lbs/ft^3 // kg/m^3
Radial shrinkage		3	%

BASSIA LATIFOLIA

Description

Found throughout practically all of the Indian peninsula, this is a tree that is cultivated more for its prized fruits and flowers —they are used to obtain oils, sugars and medicines— than for its wood, even though it has outstanding physical properties, comparable to those of teak. It reaches heights of 20 m (66 ft) at maturity. It is a slow-growing species (taking about 20 years to mature) and exhibits great variability, especially in its fruits. Its wood, which is heavy and extremely hard, has a rough surface texture and straight and uniform grain.

Beschreibung

Es handelt sich um einen Baum, der fast auf der gesamten indischen Halbinsel zu finden ist und der vielmehr wegen seiner beliebten Früchte und Blüten kultiviert wird als wegen seines Holzes. Aus den Früchten und Blüten lassen sich Öl, Zucker und Medizin herstellen. Die beachtlichen physischen Eigenschaften seines Holzes sind mit denen des Teak vergleichbar. Er ist in ausgewachsenem Zustand bis zu 20 m hoch. Es handelt sich um eine langsamwachsende Pflanze (nach 20 Jahren ausgewachsen) von großer Variabilität, besonders was seine Früchte betrifft. Sein Holz ist schwer und sehr hart. Es besitzt eine raue Oberflächentextur und eine gleichmäßige, geradlinige und gleichmäßige Maserung.

Scientific / Botanical name
Bassia latifolia

Trade / Common name
Mahua

Family name
Sapotaceae

Regions / Countries of distribution
India

Global threat status
NE

Common uses

Beams, boat construction, sleepers, building construction, wheels, turnery, heavy construction, etc.

Allgemeine Verwendung

Dachbalken, Schiffbau, Eisenbahnschwellen, Gebäudebau, Räder, Holzdrehteile, Schwerbauindustrie, etc.

Machinability

Cutting resistance	
Planing	
Polishing	

Physical properties

Numerical data	Green	Dry	English / Metric
Bending strength	9,200 // 646	13,100 // 921	psi // kgf/cm²
Impact strength	42 // 106	33 // 83	in // cm
Maximum crushing strength	4,270 // 300	7,930 // 557	psi // kgf/cm²
Stiffness	1,250 // 87	1,670 // 117	1,000 psi // 1,000 kgf/cm²
Weight	74 // 1,185	62 // 993	lbs/ft³ // kg/m³

SCHIMA WALLICHII

Description

A member of the Theaceae family, this evergreen species of a monotypic genus—commercially known as needlewood— covers a vast area of geographical distribution which includes Thailand, Malaysia, the Philippines, China and India, among other countries. When cultivated, it can reach heights as great as 27 m (90 ft) and diameters of 60 to 75 cm (24-30 in), but, in its wild state, there have been specimens with heights up to 47 m (154 ft). It produces a wood with tones ranging from grayish to reddish, with indistinguishable sapwood and heartwood, straight grain and a uniform texture. It is a dense, heavy and strong material, with mechanical properties similar to those of the teak and white oak.

Beschreibung

Schima gehört zur Familie der Teestrauchgewächse (Theaceae) und ist eine immergrüne, monotypische Pflanzengattung. Sie ist im Handel auch als Nadelholz bekannt. Geographisch ist sie auf einem weiten Gebiet verbreitet, so z. B. in Thailand, Malaysia, Philippinen, China und Indien, und anderswo. Im Anbau erreicht das Gewächs eine Höhe von höchstens 27 m und einen Stammdurchmesser von 60 bis 75 cm, aber in freier Natur sind Exemplare mit Höhen bis zu 47 m bekannt. Das Holz variiert zwischen Grau- und Rottönen und Splint- und Kernholz sind undeutlich. Es hat eine unregelmäßige Maserung aber eine gleichmäßige Struktur. Es ist ein dichtes, schweres und starkes Material, und die mechanischen Eigenschaften sind vergleichbar mit Teak oder Weiß-Eiche.

Scientific / Botanical name
Schima wallichii

Trade / Common name
Needlewood

Family name
Theaceae

Regions / Countries of distribution
Nepal, eastern India, southern China, Taiwan, and the Ryukyu Islands

Global threat status
NE

Common uses
Its wood is used for rather light products, preferably under cover: planks, plywood, beams and columns, agricultural applications, pallets, crates, turnery, toys, frames, flooring, etc.

Allgemeine Verwendung
Das Holz wird für eher leichte Produkte verwendet, insbesondere für überdachte Objekte: Platten, Sperrholz, Balken und Stützen, landwirtschaftliche Verwendung, Paletten, Kisten, Holzdrehteile, Spielzeug, Rahmen, Böden etc.

Machinability

Boring	■
Carving	■
Mortising	■
Moulding	■
Routing and recessing	■
Planing	■
Polishing	■
Sanding	■
Turning	■

Physical properties

Numerical data	Green	Dry	English / Metric
Bending strength	7,900 // 555	14,300 // 1,005	psi // kg/cm^2
Hardness		1,245 // 564	lbs // kg
Impact strength	31 // 78	27 // 68	in // cm
Maximum crushing strength	3,780 // 265	7,645 // 537	psi // kg/cm^2
Stiffness	1,360 // 95	1,970 // 138	1,000 psi // 1,000 kg/cm^2
Weight	60 // 961	41 // 656	lbs/ft^3 // kg/m^3
Radial shrinkage		5	%
Tangential shrinkage		9	%

ARBUTUS MENZIESII

Description

The Pacific madrone is a species currently in decline because of its adependence on fires to cope with the competition of the Douglas fir. Drought-resistant with a relatively fast rate of growth, it reaches heights of between 6 and 24 m (20-80 ft), with diameters ranging from 60 to 90 cm (24-36 in). Its trunk exhibits a characteristic reddish to orange tone. Although it is of little use as a construction material, its malleability makes it a highly valued raw material in furniture and marquetry industry. It is a heavy, compact and resistant wood.

Beschreibung

Der Amerikanische Erdbeerbaum ist eine schwindende Spezies, weil sie im Wettbewerb gegen die Douglasie auf Brände angewiesen ist. Er ist dürreresistent und relativ schnellwachsend. Seine Höhe liegt zwischen 6 und 24 m, sein Stammdurchmesser beträgt zwischen 60 und 90 cm. Die Farbe des Stammes ist ein charakteristisches Rot-Orange. Wenngleich sein Holz im Bereich der Baukonstruktion nur von geringem Wert ist, bewirkt seine Formbarkeit, dass es als Rohstoff im Bereich der Möbelindustrie und Intarsie sehr geschätzt wird. Das Holz ist schwer, fest und widerstandsfähig.

Scientific / Botanical name
Arbutus menziesii

Trade / Common name
Pacific madrone

Family name
Ericaceae

Regions / Countries of distribution
Western North America

Global threat status
NE

Common uses
Fuel, veneers, plywood, flooring, furniture components, interior construction, turnery, etc.

Allgemeine Verwendung
Brennstoff, Verkleidungen, Sperrholz, Böden, Möbelelemente, Innenausbau, Holzdrehteile, etc.

Machinability

Boring	
Gluing	
Mortising	
Moulding	
Nailing	
Planing	
Polishing	
Sanding	
Staining	
Steam bending	
Turning	

Physical properties

Numerical data	Green	Dry	English / Metric
Bending strength		10,400 // 731	psi // kgf/cm^2
Hardness		1,460 // 1,460	lbs // kg
Impact strength		23 // 58	in // cm
Shearing strength		1,810 // 127	psi // kgf/cm^2
Stiffness		1,230 // 86	1,000 psi // 1,000 kgf/cm^2
Weight	60 // 961	45 // 720	lbs/ft^3 // kg/m^3
Radial shrinkage		5	%
Tangential shrinkage		12	%

DIOSPYROS KAKI

Description

Known as persimmon and Chinese fig among other common names, *Diospyros kaki* is a deciduous tree native to China, India, Japan and Myanmar, but widely cultivated as an ornamental tree in subtropical to mild-temperate climates. The persimmon grows up to 6 m (19 ft) and is sometimes multi-stemmed with a round crown. Leaves are alternate, oblong, glossy on the upper surface and silky-brown underneath. The red-orange fruit is edible when ripe.

Beschreibung

Diospyros kaki ist ein sommergrüner Baum, der auch als Kakipflaume oder Persimone bekannt ist. Ursprünglich in China, Indien, Japan und Myanmar beheimatet, ist die Kultivierung als Zierpflanze in Gegenden mit subtropischem bis mildem Klima weit verbreitet. Die Persimone wächst bis zu 6 m hoch und ist manchmal mehrstämmig mit einer runden Kronenform. Ihre Blätter sind wechselständig, länglich, glänzend auf der Oberseite und samtig-braun auf der Unterseite. Die rot-orange Frucht ist in reifem Zustand genießbar.

Scientific / Botanical name
Diospyros kaki

Trade / Common name
Kaki persimmon

Family name
Ebenaceae

Regions / Countries of distribution
China

Global threat status
NE

Common uses

The wood of the persimmon is hard and heavy, black with yellowish orange, brown or gray streaks. Its close grain that takes a smooth finish is prized in Japan for fancy inlays.

Allgemeine Verwendung

Das Holz der Persimone ist hart und schwer, schwarz mit gelblichem Orange und mit braunen oder grauen Streifen. Seine feine Faserstruktur erfordert eine glatte Verarbeitung und ist in Japan als ausgefallenes Furnier gefragt.

Machinability

Boring	
Mortising	
Moulding	
Nailing	
Planing	
Polishing	

Physical properties

Numerical data	Green	Dry	English / Metric
Bending strength	10,500 // 738	18,600 // 1,307	psi // kgf/cm²
Hardness		2,300 // 1,043	lbs // kg
Maximum crushing strength	4,330 // 304	9,520 // 669	psi // kgf/cm²
Shearing strength		2,160 // 151	psi // kgf/cm²
Stiffness	1,280 // 89	1,880 // 132	1,000 psi // 1,000 kgf/cm²
Weight	63 // 1,009	52 // 832	lbs/ft³ // kg/m³
Radial shrinkage		8	%
Tangential shrinkage		11	%

DIOSPYROS VIRGINIANA

Description

The American persimmon is known (and cultivated) more for its fruit, the persimmon, than for its wood. It is a species native to North America with dimensions ranging from shrub size (5 m / 16 ft) to 30 m (100 ft), depending on the type of soil. Its bark is dark brown or dark gray, and very flaky and fragmented. Its wood, which is dark, heavy, hard and resistant with very fine grain, is actually a type of ebony. However, unlike in the case of the African or Asian ebony, the production of American persimmon wood is not viable because its heartwood is not sufficiently developed for exploitation until 100 years have passed.

Beschreibung

Die Amerikanische Persimone ist eher durch ihre Frucht, die Kaki, bekannt und wird auch wegen ihr gezüchtet als wegen ihres Holzes. Sie ist ein nordamerikanischer Endemit mit Dimensionen, die – abhängig von der Beschaffenheit des Bodens –, von der Größe eines Strauchgewächses (5 m) bis hin zu 30 m reichen. Ihre Rinde ist dunkelbraun oder dunkelgrau, bröckelig und schuppig. Ihr Holz ist dunkel, schwer, hart, widerstandsfähig und hat eine feine Maserung. Im Grunde ist es eine Art Ebenholz. Ungeachtet dessen und im Gegensatz dazu, wie mit dem Afrikanischen und Asiatischen Ebenholz vorgegangen wird, ist die Produktion von Holz im Falle der Amerikanische Persimone nicht realisierbar. Begründet liegt dies darin, dass ihr Kernholz erst nach hundert Jahren ausreichend für eine wirtschaftliche Nutzung entwickelt ist.

Scientific / Botanical name
Diospyros virginiana

Trade / Common name
American persimmon

Family name
Ebenaceae

Regions / Countries of distribution
Eastern and southern United States

Global threat status
NE

Common uses

Its scarcity makes it an extremely expensive wood that is practically used only in the production of golf clubs (woods) and veneers.

Allgemeine Verwendung

Die geringe Verfügbarkeit bewirkt, dass das Holz extrem teuer ist und praktisch nur für die Produktion von Golfschlägern oder Verkleidungen verwendet wird.

Machinability

Boring	
Gluing	
Mortising	
Nailing	
Planing	
Polishing	
Screwing	
Steam bending	
Turning	

Physical properties

Numerical data	Green	Dry	English / Metric
Bending strength	10,500 // 738	18,600 // 1,307	psi // kgf/cm^2
Hardness		2,300 // 1,043	lbs // kg
Maximum crushing strength	4,330 // 304	9,520 // 669	psi // kgf/cm^2
Shearing strength		2,160 // 151	psi // kgf/cm^2
Stiffness	1,280 // 89	1,880 // 132	1,000 psi // 1,000 kgf/cm^2
Weight	63 // 1,009	52 // 832	lbs/ft^3 // kg/m^3
Radial shrinkage		8	%
Tangential shrinkage		11	%

CARINIANA LEGALIS

Description

This species is considered the largest tree native to Brazil, as it can measure up to 50 m (164 ft) in height with diameters of up to 6 m (20 ft). They have extraordinary longevity: the oldest known specimen is about 3,000 years old. Its common name, jequitibá, can lead to confusion, as this can actually refer to different species depending on the region that one is referring to. It is a wood with a fairly uniform texture, with straight grain and a smooth and soft surface. The heartwood and sapwood barely differ and feature a color that ranges from cinnamon to brownish-pink. As a commercial wood, it has excellent properties for handling.

Beschreibung

Aufgrund seiner Höhe von bis zu 50 m und seinem Stammdurchmesser von bis zu 6 m bezeichnet man ihn als den größten Baum mit Heimat in Brasilien. Er ist außergewöhnlich langlebig: Das älteste, bekannte Exemplar ist 3 000 Jahre alt. Der allgemeine Name, Jequitiba, könnte zu Verwechslungen führen, da er sich tatsächlich auf zwei verschiedene Spezies bezieht, abhängig davon, von welcher Region die Rede ist. Das Holz hat eine sehr ebenmäßige Textur, eine geradlinige Maserung und eine glatte und geschmeidige Oberfläche. Splint- und Kernholz lassen sich kaum unterscheiden. Farblich variieren sie zwischen Zimttönen und Rotbraun. Am Markt besteht es aufgrund seiner exzellenten Eigenschaften in Bezug auf die Bearbeitung.

Scientific / Botanical name
Cariniana legalis

Trade / Common name
Jequitibá-branco

Family name
Lecythidaceae

Regions / Countries of distribution
Southeastern Brazil

Global threat status
VU

Common uses

Cabinetry, furniture, turnery, crafts, interior and exterior carpentry, flooring, plywood and veneers, etc.

Allgemeine Verwendung

Tischlerarbeiten, Mobiliar, Holzdrehteile, Kunstgewerbe, Schreinerarbeiten für die Verwendung innen und außen, Böden, Sperrholz und Verkleidungen, etc.

Machinability

Boring	
Carving	
Gluing	
Mortising	
Moulding	
Nailing	
Planing	
Routing and recessing	
Sanding	
Steam bending	
Turning	
Veneering qualities	

Physical properties

Numerical data	Green	Dry	English / Metric
Bending strength	8,244 // 579	12,907 // 907	psi // kgf/cm^2
Density		37 // 592	lbs/ft^3 // kg/m^3
Maximum crushing strength	3,916 // 275	6,397 // 449	psi // kgf/cm^2
Stiffness	1,017 // 71	1,199 // 84	1,000 psi // 1,000 kgf/cm^2
Weight	36 // 576	29 // 464	lbs/ft^3 // kg/m^3
Radial shrinkage		3.3	%
Tangential shrinkage		5.2	%

MANILKARA ZAPOTA

Description

The zapote or sapodilla is an evergreen tree native to Central America and the Caribbean but extensively cultivated today in the Philippines, India, Pakistan, Malaysia, Indonesia and Bangladesh. At maturity, its height ranges from 9 to 15 m (30-49 ft), though it is capable of growing up to 30 m (98 ft) in favorable environments. Its trunk, which exudes a resin from which gum is obtained, measures up to 50-150 cm (20-59 in). It produces a wood that is hard, heavy, resistant and exceptionally durable, practically immune to fungi and insects. With straight grain and a soft texture, its sapwood and heartwood are shaded with a backdrop of rosy or slightly reddish tones.

Beschreibung

Der Breiapfelbaum, auch Sapote, Kaugummibaum oder Sapotillbaum genannt, ist immergrün und in Mittelamerika und der Karibik zu Hause. Heute wird er extensiv auf den Philippinen, in Indien, Pakistan, Malaysia, Indonesien und Bangladesch angebaut. Ausgewachsen erreicht er eine Höhe zwischen 9 und 15 m, wobei er unter günstigen Bedingungen bis zu 30 m hoch werden kann. Sein Stamm, der zwischen 50 und 150 cm Durchmesser erreicht, sondert ein Harz ab, aus dem Kaugummi hergestellt wird. Sein Holz ist hart, schwer, widerstandsfähig und außergewöhnlich langlebig. Gegenüber Pilzen und Insekten ist es quasi resistent. Die Maserung ist geradlinig, die Textur geschmeidig. Splint- und Kernholz gehen, auf Grundlage von rosaroten Farbtönen, ineinander über.

Scientific / Botanical name
Manilkara zapota

Trade / Common name
Sapodilla

Family name
Sapotaceae

Regions / Countries of distribution
Southern Mexico, Central America, and the Caribbean

Global threat status
NE

Common uses
Its wood is barely commercialized and is available in small quantities in Florida. It is used as a construction material, and in sleepers, flooring, furniture, tool handles, etc.

Allgemeine Verwendung
Sein Holz ist kaum am Markt zu finden. Kleine Mengen sind in Florida verfügbar. Man verwendet es als Konstruktionsmaterial und in Eisenbahnschwellen, Böden, Mobiliar, Werkzeuggriffen, etc.

Machinability

Boring	
Mortising	
Moulding	
Planing	
Polishing	
Turning	

Physical properties

Numerical data	Green	Dry	English / Metric
Bending strength	18,632 // 1,309		psi // kg/cm^2
Maximum crushing strength	8,320 // 584		psi // kg/cm^2
Stiffness	2,688 // 188		1,000 psi // 1,000 kg/cm^2
Radial shrinkage		6	%
Tangential shrinkage		9	%

SASSAFRAS ALBIDUM

Description

The sassafras tree is a species native to eastern North America belonging to the Lauraceae family. The plants are normally incorporated into scrubland, with heights ranging from shrub size to as tall as 27 m (90 ft), and diameters of between 60 and 150 cm (24-60 in). As a result, in most cases the trees are not of a sufficient size to make their commercial exploitation viable; nevertheless, its wood is sold in small quantities, mixed with other species such as the ash. Its wood, which is aromatic, soft and durable, has undifferentiated sapwood and heartwood —with cinnamon tones reminiscent of the ash and chestnut— straight grain and a somewhat rough and lustrous texture.

Beschreibung

Der Sassafrasbaum ist ein nordamerikanischer Endemit, der der Familie der Lauraceae angehört. In der Regel schließen sich die Pflanzen auf Buschland zusammen. Sie erreichen Höhen, die von strauchartig bis hin zu 27 m variieren. Ihr Stamm kann einen Durchmesser zwischen 60 und 150 cm erreichen. So kommt es, dass die Bäume in den meisten Fällen die für einen kommerziellen Anbau erforderliche Größe nicht erreichen. Trotzdem wird das Holz des Sassafrasbaumes in kleinen Mengen und gemischt mit anderen Holzarten, wie beispielsweise der Esche, vertrieben. Sein aromatisches, weiches und langlebiges Holz besitzt Splint- und Kernholz. Beide sind nicht voneinander differenzierbar und in Zimttönen gehalten, die an die Esche oder die Kastanie erinnern. Die Maserung ist geradlinig, die Textur rau und glänzend.

Scientific / Botanical name
Sassafras albidum

Trade / Common name
White sassafras

Family name
Lauraceae

Regions / Countries of distribution
Eastern North America

Global threat status
NE

Common uses

Shipbuilding, foundation posts, construction materials, carpentry, sleepers, cooperage, etc.

Allgemeine Verwendung

Schiffbau, Zementierpfosten, Konstruktionsmaterial, Schreinerarbeiten, Eisenbahnschwellen, Fassbinderei, etc.

Machinability

Gluing	
Nailing	
Planing	
Screwing	

Physical properties

Numerical data	Green	Dry	English / Metric
Bending strength	6,000 // 421	9,000 // 632	psi // kg/cm^2
Maximum crushing strength	2,730 // 191	4,760 // 334	psi // kg/cm^2
Shearing strength		1,240 // 87	psi // kg/cm^2
Stiffness	910 // 63	1,120 // 78	1,000 psi // 1,000 kg/cm^2
Weight	44 // 704	31 // 496	lbs/ft^3 // kg/m^3
Radial shrinkage		4	%
Tangential shrinkage		6	%

CINNAMOMUM CAMPHORA

Description

The camphor tree is a species from the Lauraceae family that measures 18-30 m (60-100 ft) in height and between 60 and 120 cm (24-48 in) in diameter. It grows rapidly and can live more than 1,000 years. Its relevance as a commercial wood is marginal, though it does play an important role in certain niches of the market for furniture and veneers, due in large part to its malleability and the attractive texture of its wood. The sapwood and heartwood are indistinguishable from each other; they feature tones that range from pale yellow to mahogany. Before it became possible to synthesize camphor chemically, this was the species from which it was extracted.

Beschreibung

Der Kampferbaum gehört zur Familie der Lauraceae und erreicht Höhen zwischen 18-30 m und Stammdurchmesser zwischen 60 und 120 cm. Er ist schnellwachsend und gedeiht bis ins hohe Alter von tausend Jahren. Kommerziell hat sein Holz nur eine Randbedeutung, wenngleich es in bestimmten Bereichen des Marktes eine wichtige Stellung einnimmt. Seine Formbarkeit und attraktive Textur sind es, die es für die Herstellung von Mobiliar und Verkleidungen interessant machen. Splint- und Kernholz können nicht differenziert werden. Als Einheit schwanken sie farblich zwischen einem hellen Gelb und Mahagonifarben. Bevor es industriell synthetisch hergestellt werden konnte, diente die Spezies als Lieferant für Kampfer.

Scientific / Botanical name
Cinnamomum camphora

Trade / Common name
Camphor tree

Family name
Lauraceae

Regions / Countries of distribution
China, Japan, and Taiwan

Global threat status
NE

Common uses

Distillation of its wood produces camphor oil, which has antiseptic and anti-rheumatic properties. Its wood is used exclusively for veneers and furniture.

Allgemeine Verwendung

Aus dem Holz wird durch Wasserdampfdestillation Kampferöl gewonnen, das antiseptische und antirheumatische Wirkung hat. Sein Holz wird fast ausschließlich für die Herstellung von Verkleidungen und Mobiliar verwendet.

Machinability

Boring	
Moulding	
Planing	
Polishing	
Sanding	
Turning	

Physical properties

Numerical data	Green	Dry	English / Metric
Bending strength		10,165 // 714	psi // kgf/cm^2
Density		28 // 448	lbs/ft3 // kg/m^3
Maximum crushing strength		5,434 // 382	psi // kgf/cm^2
Radial shrinkage		5	%
Tangential shrinkage		8	%

OCOTEA POROSA

Description

Generally known as imbuia, *Ocotea porosa* is sometimes called Brazilian walnut, but shares no scientific similarities with trees of the *Juglans* genus and does not produce nuts. Imbuia reaches a height of 40 m (131 ft) with a trunk diameter of 1.8 m (6 ft). The base of the tree often forks and grows globular growths at the base. Leaves are long, narrow and bright green. Flowers are small, bell-shaped and yellow. The heartwood is yellowish olive to dark brown, with variegated streaks. The sapwood is gray and generally distinct. The wood of the *Ocotea porosa* is beautifully figured, hard and heavy. Imbuia is listed as a vulnerable species in the IUCN Red List.

Beschreibung

Ocotea porosa, auch bekannt als Imbuia oder Brasilianischer Nussbaum, hat keine wissenschaftliche Ähnlichkeit mit den Bäumen der Gattung *Juglans* und produziert auch keine Nüsse. Imbuia erreicht eine Höhe von 40 m und einen Stammdurchmesser von 1,8 m. Oft gabelt sich der Baum an der Basis. Zudem bildet er dort auch kugelförmige Wucherungen aus. Seine Blätter sind lang, schmal und leuchtend-grün. Die Blüten sind klein, glockenförmig und gelb. Das Kernholz ist gelblich oliv- bis dunkelbraun mit mehrfarbigen Streifen. Das Splintholz ist grau und hebt sich in der Regel ab. Das Holz von *Ocotea porosa* ist wunderschön gemustert, hart und schwer. Imbuia steht als gefährdete Art auf der Roten Liste der IUCN.

Scientific / Botanical name
Ocotea porosa

Trade / Common name
Imbuia

Family name
Lauraceae

Regions / Countries of distribution
Southern Brazil

Global threat status
VU

Common uses

Despite the fact that imbuia is listed as vulnerable, it is one of the most sought after timbers for high-quality furniture, veneers, flooring, joinery, and musical instruments.

Allgemeine Verwendung

Trotz der Tatsache, dass Imbuia als gefährdet gilt, ist es das gefragteste Holz für hochqualitative Möbel, Furniere, Bodenbeläge, Tischlerarbeiten und Musikinstrumente.

Machinability

Boring	
Gluing	
Mortising	
Moulding	
Nailing	
Planing	
Screwing	

Physical properties

Numerical data	Green	Dry	English / Metric
Weight		45 // 720	lbs/ft^3 // kg/m^3
Radial shrinkage		4.1	%
Tangential shrinkage		7.6	%

PERSEA AMERICANA

Description

Persea americana is a fast-growing tree or shrub widely known for its edible yellow-green fruit: the avocado. It is an evergreen but some varieties may loose their leaves for a short period before flowering. It grows up to 20 m (65 ft) high and develops a low, fairly dense canopy. The large, elliptical leaves are orange-red and fuzzy when young turning dark green and leathery when mature. Flowers are green and finely hairy. They occur in very large clusters but only a handful will develop into dark green and lumpy fruits the size and the shape of a pear. The fruit of the avocado is one of the most popular tropical fruits, perhaps next to kiwis.

Beschreibung

Persea americana ist ein schnellwachsender Baum oder Busch, bekannt für seine essbare, gelbgrüne Frucht: Die Avocado. Es ist ein immergrüner Baum, wobei manche Sorten ihre Blätter in der Zeit vor der Blüte verlieren können. Er wird bis zu 20 m hoch und entwickelt ein niedriges, recht dichtes Blätterdach. Die großen, elliptischen Blätter sind an jungen Exemplaren orange-rot und flaumig, an älteren Bäumen dunkelgrün und ledrig. Die Blüten sind grün und fein behaart. Sie wachsen in sehr großen Büscheln, wobei nur eine Handvoll die Entwicklung in dunkelgrüne, birnenartige Früchte vollzieht. Die Frucht der Avocado ist, neben der Kiwi, wahrscheinlich eine der bekanntesten tropischen Früchte.

Scientific / Botanical name
Persea americana

Trade / Common name
Avocado

Family name
Lauraceae

Regions / Countries of distribution
Central Mexico

Global threat status
NE

Common uses

The most valued part of the avocado is its fruit, which is edible and also used in cosmetics and traditional medicine. The small dimensions of timber available limit its use, but it is suitable for general carpentry, furniture, and paneling.

Allgemeine Verwendung

Der begehrteste Teil der Avocado ist ihre Frucht, die zum einen genießbar ist und zum anderen Einsatz in der Kosmetik und in der traditionellen Medizin findet. Aufgrund der geringen Menge an verfügbarem Holz sind auch dessen Verwendungsmöglichkeiten limitiert. Es wird allgemein für die Tischlerei empfohlen und kann für die Herstellung von Möbeln und Verkleidungen genutzt werden.

Machinability

Boring	
Mortising	
Moulding	
Planing	
Steam bending	
Veneering qualities	

Physical properties

Numerical data	Green	Dry	English / Metric
Bending strength	6,110 // 429	9,590 // 674	psi // kgf/cm^2
Density		34 // 544	lbs/ft^3 // kg/m^3
Hardness		992 // 449	lbs // kg
Maximum crushing strength	3,030 // 213	5,160 // 362	psi // kgf/cm^2
Shearing strength		885 // 62	psi // kgf/cm^2
Stiffness	1,130 // 79	1,320 // 92	1,000 psi // 1,000 kgf/cm^2
Weight	32 // 512	26 // 416	lbs/ft^3 // kg/m^3
Radial shrinkage		4	%
Tangential shrinkage		7	%

PLATANUS OCCIDENTALIS

Description

Native to the coastal forests and wetlands of the eastern United States, the occidental plane or American sycamore has also spread to Australia and Argentina as an ornamental tree, as it grows rapidly, withstands pollution and is resistant to transplantation. It grows to heights of 18 to 30 m (60-100 ft) and diameters of between 60 and 120 cm (24-36 in). The peculiar mottled appearance of its shaft is caused by the exfoliation of its bark, which peels off as the trunk grows. Although their woods have nearly identical properties, the greater resistance to infestation of the common plane (*Platanus hibrida*) favors its cultivation at the expense of its western counterpart.

Beschreibung

Ursprünglich stammt die Amerikanische Platane oder Westliche Platane aus den Feuchtgebieten und gewässernahen Wäldern des nordamerikanischen Ostens. Dennoch findet man sie stellenweise auch in Australien und Argentinien, wo sie als schnellwachsendes Zierholz eingesetzt wird, das gegenüber dem Umpflanzen und der Luftverschmutzung unempfindlich ist. Sie wächst bis auf eine Höhe von 18 bis 30 m und hat einen Durchmesser von 60-120 cm. Das charakteristisch gefleckte Aussehen ihres Stammes resultiert aus dem Abblättern ihrer Rinde, die mit Wachstum des Stammes schuppenweise abbröckelt. Obwohl beider Holz von fast identischer Qualität ist, wird die Widerstandsfähigkeit der Gemeinen Platane (*Platanus hibrida*) gegenüber Schädlingen durch ihre Kultivierung auf Kosten ihrer westlichen Artgenossen begünstigt.

Scientific / Botanical name
Platanus occidentalis

Trade / Common name
American sycamore

Family name
Platanaceae

Regions / Countries of distribution
North America

Global threat status
NE

Common uses

While not as common as its European equivalent, the wood of the occidental plane is found in the form of crates, veneers, planks and construction materials and is also used as a biomass.

Allgemeine Verwendung

Noch ist das Holz der Amerikanischen Platane nicht so verbreitet wie das ihres europäischen Pendants, man findet es jedoch in Form von Kisten, Verkleidungen, Arbeitsplatten oder Konstruktionsmaterial. Zudem wird es als Biomasse eingesetzt.

Machinability

Boring	
Gluing	
Mortising	
Moulding	
Nailing	
Planing	
Polishing	
Screwing	
Staining	
Steam bending	
Turning	
Varnishing	

Physical properties

Numerical data	Green	Dry	English / Metric
Bending strength	6,500 // 456	10,000 // 703	psi // kgf/cm^2
Hardness		770 // 349	lbs // kg
Impact strength	26 // 66	26 // 66	in // cm
Maximum crushing strength	2,920 // 205	5,380 // 378	psi // kgf/cm^2
Shearing strength		1,470 // 103	psi // kgf/cm^2
Stiffness	1,060 // 74	1,420 // 99	1,000 psi // 1,000 kgf/cm^2
Weight	52 // 832	34 // 544	lbs/ft^3 // kg/m^3
Radial shrinkage		5	%
Tangential shrinkage		8	%

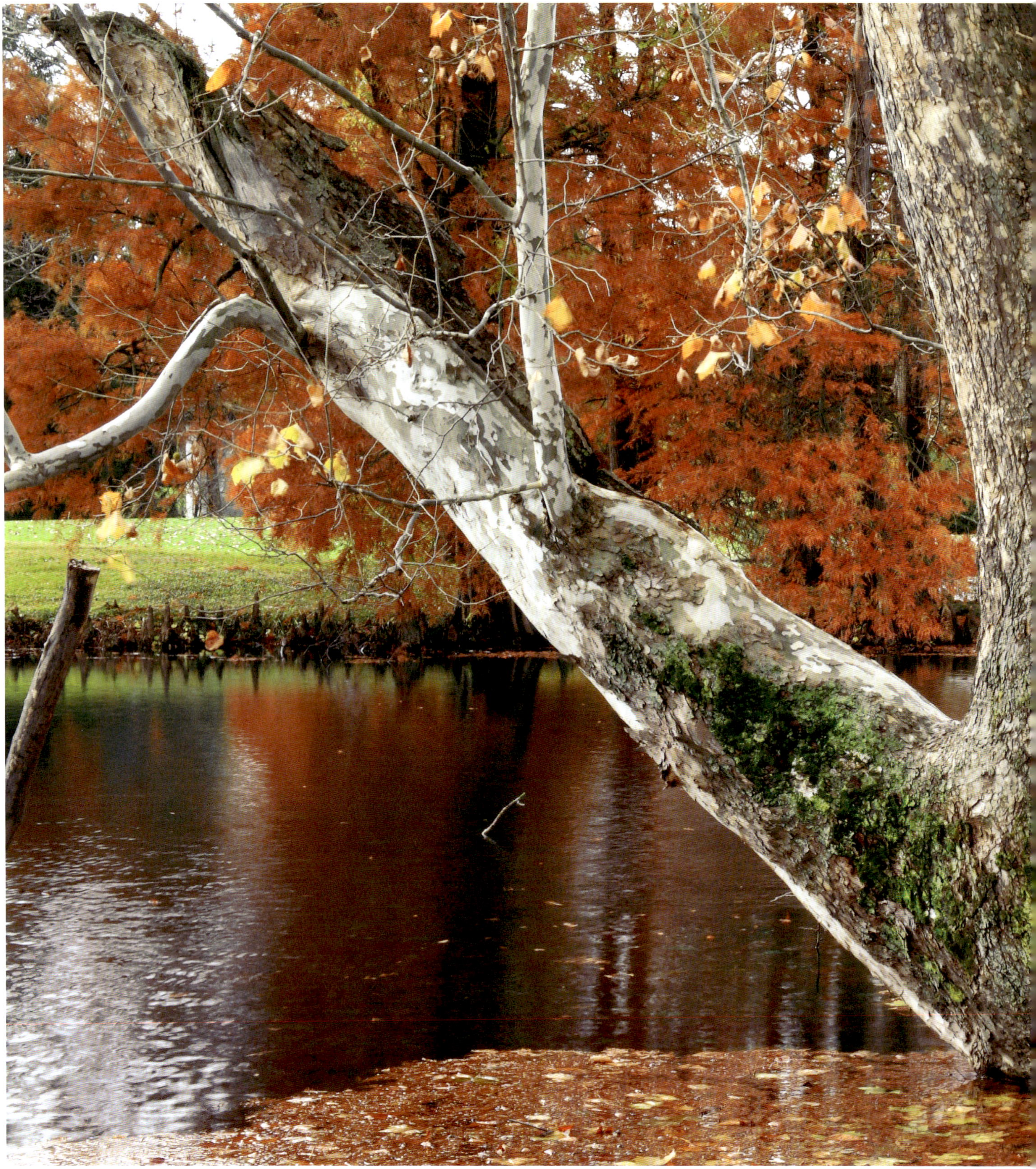

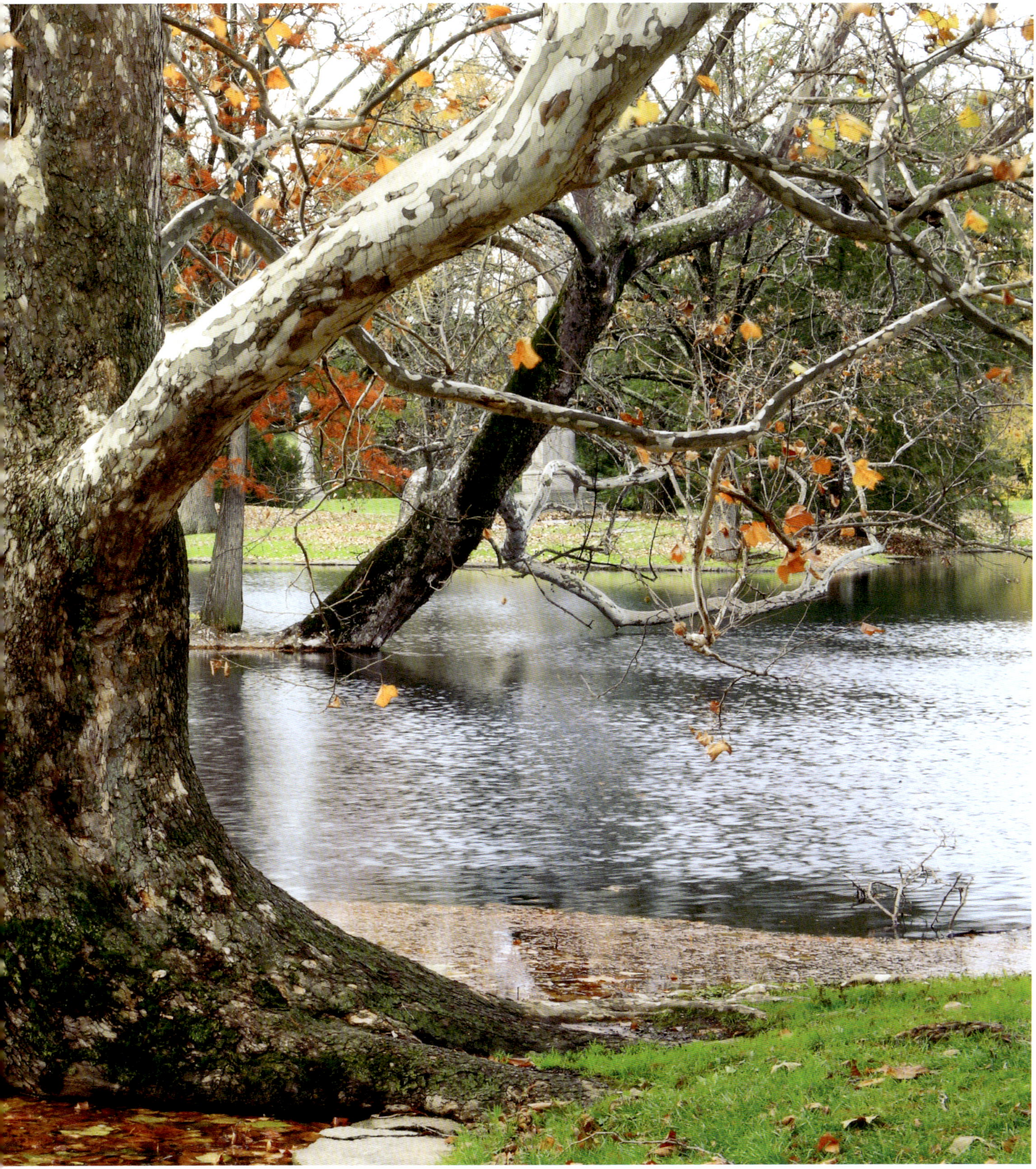

GREVILLEA ROBUSTA

Description

The silky oak or Australian oak, native to the east coast of Australia, is the largest member of the genus *Grevillea*. It is one of the most rapidly growing trees in existence and is, in general, very popular as a tree for lining roads, and is used as a hedge and a landscaping tree. Its wood is heavily commercialized in Australia, even though its harvest is restricted, but it is rare overseas, given the associated transport and processing costs, which preclude it from being competitive outside of its natural market. It is a porous and very lustrous wood, with typically straight grain and dotted with reddish tones on a background the color of cinnamon, and possessing physical properties that make it ideal for veneers.

Beschreibung

Die Australische Silbereiche stammt von der Ostküste Australiens und ist der größte Exponent der Gattung der *Grevillea*. Sie ist einer der schnellwachsensten Bäume, die es gibt und wird in der Regel bevorzugt als Straßensäumung, Hecke oder im Landschaftsbau eingesetzt. Trotz einer Beschränkung der Abholzung ist ihr Holz auf dem australischen Markt sehr präsent. Übersee hingegen findet man es nur selten, da die Kosten für den Transport und damit verbundene Prozesse eine Konkurrenzfähigkeit außerhalb ihres natürlichen Marktes verhindern. Das Holz ist porös, sehr glänzend und hat typischerweise eine geradlinige Maserung mit rötlichen Flecken auf zimtfarbenem Grund - ästhetische Eigenschaften, die optimale Voraussetzungen für eine Verwendung als Verkleidungsmaterial darstellen.

Scientific / Botanical name
Grevillea robusta

Trade / Common name
Southern silky oak

Family name
Proteaceae

Regions / Countries of distribution
Eastern Australia

Global threat status
NE

Common uses

Now replaced by aluminum, it was formerly used for window frames because of its weather resistance. It is found in musical instruments, cabinetry, moldings, furniture, balustrades, flooring, etc.

Allgemeine Verwendung

Heute durch Aluminium ersetzt, wurde sie früher, aufgrund ihrer Witterungsbeständigkeit, für die Herstellung von Fensterrahmen eingesetzt. Man findet ihr Holz zudem in Form von Musikinstrumenten, Tischlerarbeiten, Leisten, Mobiliar, Balustraden, Böden, etc.

Machinability

Boring	
Carving	
Mortising	
Moulding	
Nailing	
Planing	
Polishing	
Sanding	
Staining	
Turning	

Physical properties

Numerical data	Green	Dry	English / Metric
Bending strength		10,930 // 768	psi // kgf/cm^2
Hardness		840 // 381	lbs // kg
Maximum crushing strength		5,060 // 355	psi // kgf/cm^2
Stiffness		1,110 // 78	1,000 psi // 1,000 kgf/cm^2
Weight		38 // 608	lbs/ft^3 // kg/m^3
Radial shrinkage		3	%
Tangential shrinkage		8	%

PLATANUS HYBRIDA

Description

What we now know as the common plane is actually a hybrid between the *Platanus orientalis*, propagated by Greeks and Romans in Europe and the Middle East, and the *Platanus occidentalis*, native to the Atlantic coast of North America. This fact accounts for its extraordinary variability, which often makes its classification difficult. It is a large species —it measures up to 30 or 40 m (100-130 ft) in height and about 90-120 cm (36-48 in) in diameter— having a relatively long lifespan —the oldest known specimens have lived for 300 years— and growing extraordinarily rapidly. Its wood has a texture dominated by wide, brownish gray stripes, which contrast with the light background.

Beschreibung

Die Spezies, die heute als Gemeine Platane oder Ahornblattrige Platane bezeichnet wird, ist in Wirklichkeit eine Kreuzung aus *Platanus orientalis* und *Platanus occidentalis*. *Platanus orientalis* wurde durch die Griechen und die Römer in Europa und dem Nahen Osten verbreitet, *Platanus occidentalis* ist ursprünglich auf der atlantischen Ebene Nordamerikas heimisch. Dies ist eine Gegebenheit, die ihre enorme Wandelbarkeit unterstreicht, durch die eine Klassifikation oftmals erschwert wird. Es ist eine außergewöhnlich schnellwachsende Spezies von stattlicher Größe, die eine Höhe zwischen 30 und 40 m erreicht und deren Stammdurchmesser zwischen 90 und 120 cm beträgt. Ihre ältesten bekannten Exemplare sind knapp drei Jahrhunderte alt, was die Gemeine Platane zu einer relativ langlebigen Spezies macht. Die Textur ihres Holzes wird von breiten, braunen Linien dominiert, die sich von einem hellen Grund abheben.

Scientific / Botanical name
Platanus hybrida

Trade / Common name
Hybrid plane

Family name
Platanaceae

Regions / Countries of distribution
Cultivated origin

Global threat status
NE

Common uses

The attractive texture of its wood makes it a suitable material for plywood, panels and veneers. Though less common, it's sometimes used in cabinetry or as a carving wood.

Allgemeine Verwendung

Die attraktive Textur ihres Holzes macht sie zu geeignetem Material für Sperrholz, Paneele und Verkleidungen. Obgleich weniger gebräuchlich wird es nicht selten für Tischlerarbeiten und Holzschnitzerei eingesetzt.

Machinability

Boring	
Gluing	
Mortising	
Nailing	
Planing	
Polishing	
Staining	
Steam bending	
Turning	

Physical properties

Numerical data	Green	Dry	English / Metric
Bending strength	7,505 // 527		psi // kgf/cm²
Hardness		1,270 // 576	lbs // kg
Maximum crushing strength	3,370 // 236	5,798 // 407	psi // kgf/cm²
Stiffness	995 // 69		1,000 psi // 1,000 kgf/cm²
Weight	49 // 784	40 // 640	lbs/ft³ // kg/m³

FAUREA SALIGNA

Description

The generic name, *Faurea*, was given to this tree as a homage to a young soldier and botanist, W. C. Faure. The specific name, *saligna*, means "willow-like" in Latin and makes reference to the droopy leaves. *Faurea saligna* or willow beechwood is a semi-deciduous tree growing to 7-10 m (22-32 ft) in height with a sparse spreading crown. The trunk can be straight or twisted with a diameter of around 60 cm (23 in). Its bark is dark gray-brown and deeply fissured. Leaves are alternate, glossy, yellowish green, turning red in the fall adding grace to the tree. Flowers are greenish white during the flowering phase and turn purple or light pink when mature. The willow beechwood produces small club-shaped brown nuts with long silky hairs.

Beschreibung

Der Gattungsname, *Faurea*, wurde dem Baum in Erinnerung an den jungen Soldaten und Botaniker W. C. Faure gegeben. Der Artname, *saligna*, bedeutet aus dem Lateinischen „weidenartig" und bezieht sich auf die herabhängenden Blätter. *Faurea saligna* ist ein halbimmergrüner Baum, der 7-10 m hoch wächst und eine spärlich ausgebreitete Krone besitzt. Der Stamm kann bei einem Durchmesser von 60 cm gerade oder gewunden sein. Die Rinde ist dunkelgraubraun mit tiefen Furchen. Die Blätter sind wechselständig, glänzend, gelblich-grün. Wenn sie im Herbst rot werden, verleihen sie dem Baum Anmut. Während der Blütezeit sind die Blüten grünlich-weiß, werden dann jedoch lila oder hellpink. *Faurea saligna* produziert kleine, keulenförmige Nüsse mit langen, seidigen Haaren.

Scientific / Botanical name
Faurea saligna

Trade / Common name
Willow beechwood

Family name
Proteaceae

Regions / Countries of distribution
Tropical Africa

Global threat status
NE

Common uses

The timber of the willow beechwood is easy to work and good for furniture, joinery, and paneling. Soaking the wood in water produces a red dye and the bark can be used for tanning leather.

Allgemeine Verwendung

Das Holz ist leicht bearbeitbar und für die Herstellung von Möbeln, Tischlereiarbeiten und Verkleidungen geeignet. Weicht man das Holz in Wasser ein, entsteht eine rote Farbe. Die Rinde kann zudem für das Gerben von Leder verwendet werden.

Machinability

Nailing	
Planing	
Polishing	
Turning	

Physical properties

Numerical data	Green	Dry	English / Metric
Bending strength	8,245 // 579	12,905 // 907	psi // kgf/cm²
Density		50 // 800	lbs/ft³ // kg/m³
Hardness		1,660 // 752	lbs // kg
Maximum crushing strength	5,900 // 414	8,760 // 615	psi // kgf/cm²
Shearing strength		1,990 // 139	psi // kgf/cm²
Stiffness	1,450 // 101	1,680 // 118	1,000 psi // 1,000 kgf/cm²
Weight	48 // 768	40 // 640	lbs/ft³ // kg/m³
Radial shrinkage		5	%
Tangential shrinkage		8	%

NYSSA SYLVATICA

Description

The black tupelo is a monoecious deciduous species native to eastern North America, though it is also found in isolated pockets in Mexico. In the right conditions —ideally, those existing in landscapes with small, cleared hillsides and terraces in the southeastern United States— it grows to 20-25 m (66-82 ft), or, in rare cases, up to 35 m (115 ft). Its trunk, with dark gray bark that is very flaky in young specimens and full of cracks at maturity, usually measures 60-90 cm (24-36 in). Its wood is rot-prone but responds very well to preservatives. It is commonly used as a timber-producing species in commercial markets in the United States, where it competes with other low-level woods.

Beschreibung

Der Wald-Tupelobaum ist eine monözische, laubabwerfende und endemische Spezies aus dem Osten Nordamerikas, wenngleich sie auch in entlegenen Gegenden Mexikos zu finden ist. Unter geeigneten Bedingungen wie den niedrigen, hellen Hängen und Terrassen des Südostens der Vereinigten Staaten kann er eine Höhe zwischen 20 und 25 m, in seltenen Fällen sogar von bis zu 35 m, erreichen. Sein Stamm hat eine dunkelgraue Rinde, die an jungen Exemplaren schuppig, an älteren sehr rissig ist. Für gewöhnlich hat er einen Durchmesser von 60-90 cm. Sein Holz ist fäuleanfällig, nimmt jedoch Schutzanstriche gut an. Das Nutzholz dieser Spezies ist in der Regel auf den Handelswegen der Vereinigten Staaten zu finden, wo es mit anderen Holzarten niederen Ranges konkurriert.

Scientific / Botanical name
Nyssa sylvatica

Trade / Common name
Black tupelo

Family name
Cornaceae

Regions / Countries of distribution
Eastern North America

Global threat status
NE

Common uses

Its wood is somewhat instable and is difficult to split. Thus, it is typically found in the form of wood pulp, pallets, pulleys, and sleepers or, less commonly, as a material for flooring and furniture.

Allgemeine Verwendung

Das Holz ist leicht und instabil und kann mühelos zerlegt werden. Daher kommt es, dass es für gewöhnlich in Form von Zellstoff, Paletten, Seilrollen, Eisenbahnschwellen oder, in selteneren Fällen, als Material für Böden und Mobiliar vorzufinden ist.

Machinability

Boring	🟧
Gluing	🟧
Mortising	🟨
Moulding	🟧
Nailing	🟧
Planing	🟧
Sanding	🟨
Screwing	🟧
Steam bending	🟨
Turning	🟧

Physical properties

Numerical data	Green	Dry	English / Metric
Bending strength	7,000 // 492	9,600 // 674	psi // kg/cm²
Hardness		810 // 367	lbs // kg
Impact strength	30 // 76	22 // 55	in // cm
Maximum crushing strength	3,040 // 213	5,520 // 388	psi // kg/cm²
Shearing strength		1,340 // 94	psi // kg/cm²
Stiffness	1,030 // 72	1,200 // 84	1,000 psi // 1,000 kg/cm²
Weight	45 // 720	35 // 560	lbs/ft³ // kg/m³
Radial shrinkage		5	%
Tangential shrinkage		9	%

CORNUS FLORIDA

Description

The flowering dogwood is a small deciduous tree of the family Cornaceae that reaches heights of about 9 m (30 ft), and diameters of approximately 20 cm (8 in). The wood of the dogwood, which is very scarce —so much so that it is sometimes sold by the pound— and 90% of which is obtained from the sapwood, is among the most expensive. It has a compact, soft and lustrous texture, with closed pores and a highly interwoven grain. The sapwood features reddish brown tones; the heartwood, which is highly concentrated and uneven, is dark brown. It is an extraordinarily dense and heavy wood, with properties that are vastly superior to those of the teak or maple.

Beschreibung

Der Blüten-Hartriegel, auch Amerikanischer Blumen-Hartriegel genannt, ist ein kleiner, laubabwerfender Baum aus der Familie der Cornaceae, der um die 9 m Höhe und einen Stammdurchmesser von etwa 20 cm aufweist. Das Holz des Blüten-Hartriegels, das zu 90% aus seinem Splintholz gewonnen wird, ist äußerst knapp und steht auf der Liste der teuersten Holzsorten. Es ist von fester, geschmeidiger und glänzender Textur, mit geschlossenen Poren und sehr verflochtener Maserung. Das Splintholz ist in Rotbrauntönen gehalten, das Kernholz, sehr konzentriert und uneinheitlich, hat ein dunkles Kastanienbraun. Das Holz ist außergewöhnlich dicht und schwer. Seine Eigenschaften überragen die des Teak oder des Ahorns weitestgehend.

Scientific / Botanical name
Cornus florida

Trade / Common name
Flowering dogwood

Family name
Cornaceae

Regions / Countries of distribution
Eastern North America and eastern Mexico

Global threat status
NE

Common uses

Textile equipment, easels, sports equipment, bobbins and spools, levers, pulleys, etc.

Allgemeine Verwendung

Textilausstattungen, Gestelle, Sportartikel, Rollen und Spulen, Hebel, Scheiben, etc.

Machinability

Gluing	
Planing	
Polishing	
Turning	

Physical properties

Numerical data	Green	Dry	English / Metric
Bending strength	9,500 // 667	18,917 // 1,329	psi // kgf/cm^2
Impact strength	28 // 71		in // cm
Maximum crushing strength	3,640 // 255	9,985 // 702	psi // kgf/cm^2
Stiffness	1,410 // 99	2,361 // 165	1,000 psi // 1,000 kgf/cm^2
Weight	64 // 1,025	51 // 816	lbs/ft^3 // kg/m^3
Radial shrinkage		7	%
Tangential shrinkage		12	%

LIQUIDAMBAR STYRACIFLUA

Description

The American sweetgum is native to the temperate regions of Mexico and the United States, but it now grows on nearly every continent, due to the attractiveness of its intense autumn foliage. In its wild state, it reaches heights of between 18 and 30 m (60-100 ft) and diameters of between 50 and 90 cm (18-36 in). Its sapwood, which is easily penetrated by the dyes used to simulate other woods —cherry, oak, maple or mahogany—, is marketed as an affordable panel or surface; its heartwood —with a mahogany background covered by dark resinous veins, which, along with its satiny surface, give it a marbled appearance—, on the other hand, is one of the most expensive and popular interior woods on the European market.

Beschreibung

Der Amerikanische Amberbaum ist ursprünglich in den gemäßigten Gebieten Mexikos und der Vereinigten Staaten beheimatet. Dank seiner attraktiven Herbstfärbung kommt er heute als Zierholz auf fast allen Kontinenten vor. Wild wachsende Exemplare erreichen eine Höhe von 18-30 m und einen Stammdurchmesser zwischen 50 und 90 cm. Sein Splintholz ist sehr aufnahmefähig was Farbe betrifft, die man benutzt, um andere Holzsorten, wie Kirsche, Eiche, Ahorn oder Mahagoni zu simulieren. Es wird als Paneel oder Verkleidung zu erschwinglichen Preisen vertrieben. Sein Kernholz hingegen gehört zu den teuersten und gefragtesten Holzsorten auf dem europäischen Markt. Sein mahagonifarbener Grundton durchzogen mit dunklen Harzadern verleiht dem Holz, in Verbindung mit seiner satinierten Oberfläche, ein marmorhaftes Aussehen.

Scientific / Botanical name
Liquidambar styraciflua

Trade / Common name
American sweetgum

Family name
Altingiaceae

Regions / Countries of distribution
Eastern North America

Global threat status
LC

Common uses

Because it is a soft wood with limited resistance and durability, its use is limited to interior decorative items, with the exception of its heartwood, which is common in cabinetry.

Allgemeine Verwendung

Da es sich um ein weiches Holz mit geringer Widerstandsfähigkeit und Langlebigkeit handelt, beschränkt sich seine Verwendbarkeit auf dekorative Innenelemente. Ausnahme bildet unter Umständen das Kernholz, das als Material für Tischlerarbeiten etabliert ist.

Machinability

Boring	
Gluing	
Mortising	
Moulding	
Nailing	
Planing	
Polishing	
Sanding	
Staining	
Steam bending	
Turning	

Physical properties

Numerical data	Green	Dry	English / Metric
Bending strength	7,100 // 499	12,500 // 878	psi // kgf/cm^2
Density		35 // 560	lbs/ft^3 // kg/m^3
Hardness		850 // 385	lbs // kg
Impact strength	36 // 91	32 // 81	in // cm
Maximum crushing strength	3,040 // 213	6,320 // 444	psi // kgf/cm^2
Shearing strength		1,100 // 77	psi // kgf/cm^2
Stiffness	1,200 // 84	1,640 // 115	1,000 psi // 1,000 kgf/cm^2
Weight	55 // 881	35 // 560	lbs/ft^3 // kg/m^3
Radial shrinkage		6	%
Tangential shrinkage		10	%

CERCIDIPHYLLUM JAPONICUM

Description

The katsura is a species native to Japan and eastern China that is widely cultivated as an ornamental tree in humid areas for the vibrant autumn colors of its foliage and also emits a unique fragrance resembling burnt sugar. In its wild state, it reaches heights of up to 45 m (148 ft) and diameters of about 2 m (79 in) —making it one of the largest species in its habitat— but when cultivated or in dry conditions, it rarely exceeds 15 m (49 ft). Its wood, which has a fine texture and is diffuse-porous, with sapwood and heartwood exhibiting uniform brown tones and interwoven grain, is rather soft and light, which makes it especially useful as a material for the processing industries.

Beschreibung

Der Japanische Kuchenbaum, auch Japanischer Katsurabaum genannt, ist ein Endemit, der aus Japan und dem Osten Chinas stammt. Wegen seiner lebendigen Herbstfärbung wird er in feuchten Gegenden vermehrt als Zierholz kultiviert, das zudem einen charakteristischen Duft nach verbranntem Zucker absondert. Wild wachsend erreicht der Japanische Kuchenbaum eine Höhe von bis zu 45 m und einen Stammdurchmesser von etwa 2 m, was ihn zu einem der größten Bäume seines Habitats macht. Unter Zuchtbedingungen überschreitet er allerdings selten eine Höhe von 15 m. Sein Holz ist von einer zarten Textur mit ausgedehnten Poren, nicht differenzierbarem Splint- und Kernholz in Brauntönen und verflochtener Maserung. Es ist eher leicht und weich und damit ideal als Material für die Verarbeitungsindustrie.

Scientific / Botanical name
Cercidiphyllum japonicum

Trade / Common name
Katsura

Family name
Cercidiphyllaceae

Regions / Countries of distribution
China, and Japan

Global threat status
NT

Common uses

It serves mainly as an ornamental tree, but, as a timber-yielding species, it is highly valued as a material for furniture and interior finishes.

Allgemeine Verwendung

Grundsätzlich dient er als Ziergehölz. Als Nutzholz ist er beliebtes Material für die Bereiche Mobiliar und Innenausbau.

Machinability

Mortising	
Moulding	
Painting	
Polishing	
Veneering qualities	

Physical properties

Numerical data	Green	Dry	English / Metric
Density		25 // 400	lbs/ft³ // kg/m³
Weight	25 // 400	21 // 336	lbs/ft³ // kg/m³
Radial shrinkage		4.1	%
Tangential shrinkage		6	%

BALANITES AEGYPTIACA

Description

The desert date is a medium-sized deciduous tree —reaching no more than 8 m (26 ft) in height— which is very prickly, with convoluted branches and deep roots. Its shaft is well defined or slightly twisted and its brownish gray bark is thick and cracked. Its fruit is a species of date or fleshy drupe that is olive green in color. It has a very slow growth cycle, which offsets its adaptability to extreme droughts and fires. It is used mainly for fuel, fertilizer and fodder; its use as a construction material has historically been residual.
Its wood, however, is heavy, durable and malleable, which makes it suitable as a raw material for furniture and utensils.

Beschreibung

Die Wüstendattel ist ein mittelgroßer, laubabwerfender Baum, der eine Höhe von 8 m nicht überschreitet. Er ist sehr stachelig, dicht verzweigt und tief verwurzelt. Sein Stamm ist definiert oder leicht verdreht und seine Rinde von graubrauner Färbung, grob und rissig. Seine Frucht ist eine Art Dattel oder fleischige, olivfarbene Steinfrucht. Der sehr langsame Wachstumszyklus kompensiert die Anpassungsfähigkeit an extreme klimatische Bedingungen wie Dürre und Brände. Die hauptsächliche Verwendung erfolgt als Brennstoff, Dünger oder Futtermittel. Die Verwendung als Konstruktionsmaterial ist aus der Vergangenheit übrig geblieben. Sein Holz jedoch ist schwer, langlebig und formbar und macht es zum idealen Rohstoff für Mobiliar und Werkzeuge.

Scientific / Botanical name
Balanites aegyptiaca

Trade / Common name
Desert date tree

Family name
Zygophyllaceae

Regions / Countries of distribution
Africa, and Middle East

Global threat status
NE

Common uses
Furniture, kitchen utensils, farm tools, carpentry, etc.

Allgemeine Verwendung
Mobiliar, Küchengeräte, landwirtschaftliches Gerät, Schreinerarbeiten, etc.

Machinability

Boring	
Gluing	
Mortising	
Moulding	
Nailing	
Planing	
Varnishing	

Physical properties

Numerical data	Green	Dry	English / Metric
Bending strength	9,310 // 654	14,566 // 1,024	psi // kgf/cm²
Density		47 // 752	lbs/ft³ // kg/m³
Hardness		2,318 // 1,051	lbs // kg
Impact strength		47 // 119	in // cm
Maximum crushing strength	3,916 // 275	6,397 // 449	psi // kgf/cm²
Shearing strength		2,462 // 173	psi // kgf/cm²
Stiffness	1,347 // 94	1,563 // 109	1,000 psi // 1,000 kgf/cm²
Weight	46 // 736	37 // 592	lbs/ft³ // kg/m³
Radial shrinkage	2		%
Tangential shrinkage	5		%

GUAIACUM OFFICINALE

Description

The roughbark *lignum-vitae* is an evergreen that grows to a height of 9-12 m (29.5-39 ft) with a crooked trunk, knotty branches and a dense crown, which for a good part of the year is covered by abundant blue flowers. Leaves are pinnate, oval obtuse and fruit is an obcordate capsule. The color of the wood ranges from pale yellow-green to deep green or brown, darkening with age, especially when exposed to light. The grain is usually interlocked.

The trade of the roughbark *lignum-vitae* is restricted since it has been exploited to the brink of extinction and when available, its source is often questionable. Consequently, prices are very high especially since the timber is sold by the pound.

Beschreibung

Lignum-vitae ist ein immergrüner Baum mit einer Höhe von 9-12 m, einem gekrümmten Stamm, knotigen Ästen und einer dichten Krone, die die meiste Zeit des Jahres über mit opulenten, blauen Blüten bedeckt ist. Die Blätter sind gefiedert, abgestumpft-oval, die Frucht ist eine herzförmige Kapsel. Die Farbe des Holzes reicht von einem hellen Gelbgrün bis hin zu einem Dunkelgrün oder Braun. Mit zunehmendem Alter oder wenn es dem Licht ausgesetzt ist, dunkelt es nach. Die Maserung ist verzahnt.

Seit einer starken Ausbeutung, die den Baum an den Rand des Aussterbens brachte, ist Handel mit *lignum-vitae* nur noch begrenzt erlaubt. Wenn er verfügbar ist, dann meist von einer fraglichen Quelle. Dementsprechend ist der Holzpreis sehr hoch, vor allem seit es pfundweise verkauft wird.

Scientific / Botanical name
Guaiacum officinale

Trade / Common name
Roughbark *lignum-vitae*

Family name
Zygophyllaceae

Regions / Countries of distribution
Caribbean and northern coast of South America

Global threat status
EN

Common uses

The name *lignum-vitae* means "tree of life" or "wood of life" in Latin, which derives from the tree's many medicinal uses. The wood is among the densest and heaviest, making it an invaluable construction material.

Allgemeine Verwendung

Der Name *Lignum-Vitae* leitet sich von seiner Verwendung in der Medizin ab und bedeutet auf Latein „Baum des Lebens" oder „Holz des Lebens". Als eine der dichtesten und schwersten Holzarten, ist es als Konstruktionsmaterial von unschätzbarem Wert.

Machinability

- Boring
- Carving
- Gluing
- Mortising
- Moulding
- Planing
- Polishing
- Routing and recessing
- Sanding
- Splitting
- Turning

Physical properties

Numerical data	Green	Dry	English / Metric
Density		79 // 1,265	lbs/ft^3 // kg/m^3
Hardness		4,500 // 2,041	lbs // kg
Maximum crushing strength		11,400 // 801	psi // kgf/cm^2
Weight		84 // 1,345	lbs/ft^3 // kg/m^3
Radial shrinkage		5	%
Tangential shrinkage		8	%

CREDITS

All wood grain images were kindly provided by the Xiloteca Manuel Soler (www.xiloteca.com).

© a.dombrowski
p. 162 left

© Abu Shawka
p. 346

© aha
p. 302 upper left

© Akos Kokai
pp. 182-183

© Aleksander Bolbot
pp. 392-393

© Alois Staudacher
p. 44 left

© Alvesgaspar
p. 388 upper and lower left

© Andrea J. Smith
pp. 136-137

© Androstachys
p. 314 right

© Anton Gvozdikov
pp. 358-359

© Aresauburn
pp. 478-479

© Arkorn
pp. 202-203

© Arthur Chapman
p. 106 left

© Arturo Reina
p. 192 right

© asife
pp. 318-319

© Atamari
pp. 254 lower left and right, 374 right, 400 lower left and right

© B.navez
p. 448

© Barry Stock
p. 404

© Benjamin Gimmel
p. 336 upper left

© Bill Cook
pp. 64 right, 142 right

© BMJ
pp. 92-93

© Bobistraveling
pp. 138 left, 242 upper left

© Böhringer Friedrich
p. 462 left

© Bonnie Watton
pp. 244-245

© Botaurus
p. 94 right

© brewbooks
pp. 460-461

© Bri Weldon
p. 160 right

© Brian Erickson
pp. 74-75

© Bruce Marlin
pp. 278 left, 454 left

© Bruce McAdam
p. 142 left

© Carla Antonini
p. 254 upper left

© Carla Van Wagoner
pp. 452-453

© Chhe
pp. 224 right, 230 right, 316 right, 324 left, 432 right, 468 right

© Chris Hellyar
pp. 194-195

© Chris M
pp. 56-57

© Chris M Morris
pp. 470-471

© Chris Waits
pp. 362-363

© Christoph Neumüller
p. 342 right

© Cindy Sims Parr
pp. 222-223

© Clinton Steeds
p. 84 upper left

© Collpicto
pp. 116-117

© Concobongo1041
p. 340 left

© Cousin_Avi
pp. 306-307

© Crusier
pp. 14 right, 36 left, 80 left, 204 left, 484-485

© D and D Photo Sudbury
pp. 128-129

© Daderot
pp. 258 left, 266 left, 350 left, 384 left, 428 right, 446

© Dalgial
pp. 328 upper left, 366, 454 right

© dalvenjah
p. 192 left

© Daniel Mayer
p. 76 left

© Daniel Prudek
pp. 338-339, 390-391

© Darkone
p. 286 left

© Dave Powell, USDA Forest Service
p. 130 right

© David Sankbone
p. 274 right

© Dawn J Benko
pp. 140-141

© Dawn Endico
p. 50 left

© Dcrjsr
p. 196 right

© Dean Pennala
pp. 124-125

© Denis Blofield
pp. 104-105

© Denton Rumsey
pp. 152-153

© Dereck Ramsey
pp. 90 right, 242 right, 278 right, 472 right, 476 right, 480 right

© DM
pp. 426-427

© Doug Lemke
pp. 206-207, 456-457

© Dr. Morley Read
pp. 376-377

© eastvanfran
pp. 418-419

© eddie104
pp. 444-445

© elwynn
pp. 260-261

© Evgeniya Uvarova
pp. 464-465

© Eye-blink
pp. 20-21

© Fanghong
p. 368 left

© Forest & Kim Starr
pp. 22, 160 left, 210, 246, 270, 308 left, 354 right, 372 left, 394 left, 406 left, 408 left, 458 left, 488

© Fotomine
pp. 208-209

© Frank Kovalchek
pp. 40-41

© Gagea
p. 118 right

© Gaspar Avila
pp. 16-17

© Geographer
pp. 314 left, 364 left

© Gertjan Hooijer
pp. 48-49

© H. Zell
pp. 26 right, 350 right

© Hannes Grobe
p. 374 left

© Harvey Barrison
p. 26 left

© Hector Garcia Serrano
pp. 402-403

© Hedwig Storch
p. 42 right

© Holbox
pp. 212-213

© Iftahm
p. 248 lower left

© Ilya Katsnelson
p. 84 lower left

© ISAKA Yoji
p. 274 lower left

© Isfisk
p. 224 left

© J Brew
p. 184 left

© J.M. Garg
p. 412 right

© JMK
pp. 282 right, 466 left

© James Steakley
p. 436 right

© Jami Dwyer
p. 70 right

© Jane Shelby Richardson
p. 66 right

© Jaroslav Machacek
pp. 6-7

© Jason Patrick Ross
pp. 164-165

© jayeshpatil912
p. 236 right

© Jean-Pol GRANDMONT
pp. 110 left, 242 right, 274 upper left, 388 lower left, 468 left, 480 left

© Jen duMoulin
pp. 304-305

© Jennifurr-Jinx
p. 114 left

© Jerzy Strzelecki
p. 188

© Ji-Elle
p. 486 right

© Jiri Hera
pp. 386-387

© João Medeiros
p. 308 right

© John C. Hooten
pp. 352-353

© John J. O'Brien
p. 50 right

© John Lindsay-Smith
pp. 348-349

© Joloei
pp. 450 451

© Jon Richfield
p. 300 lower left

© Josemanuel
p. 342 lower left

© Joseph O'Brien, USDA Forest Service
p. 120 left

© Joshua Mayer
pp. 122-123

© jps
pp. 344-345

© Jutta234
p. 380 right

© Kahuroa
pp. 24 upper left and right, 174, 180

© Karduelis
p. 422 lower left

© Karora
p. 24 lower left

© Keith Kanoti, Maine Forest Service, Bugwood.org
pp. 54 left, 106 right, 126

© Ken Schulze
pp. 382-383

© Kennerth Kullman
pp. 38-39

© KENPEI
pp. 138 right, 168 right, 200 right, 286 right, 328 lower left, 364 right, 368 right, 414 left, 422 upper left, 440 left

© Kletr
pp. 144-145

© krugergirl26
pp. 252-253

© Krzysztof P. Jasiutowicz
p. 214 lower left

© Lauren Gutierrez
pp. 394 right, 396-397, 398-399

© Lijuan Guo
pp. 482-483

© Linda Baird-White
p. 134 left

© Liné1
pp. 214 upper left, 262 right, 328 right, 422 right, 472 upper left

© liveostockimages
pp. 290-291

© Louise Wolff
p. 220 left

© Luis Fernández García
p. 228 lower left

© Luri
pp. 158-159

© Lusitana
p. 42 left

© M. Cornelius
pp. 98-99

© M. Shcherbyna
pp. 234-235

© Magnus Manske
p. 424 left

© Malachi Jacobs
pp. 474-475

© Maljalen
pp. 370-371

© Manfred Heyde
p. 302 lower left and right

© Manuel.flury
p. 400 upper left

© Marco Schmidt
pp. 272, 372 right, 486 left

© Marehogomcoy
p. 266 right

© Maren Wulf
pp. 378-379

© Mariense
p. 168 left

© Marion Schneider & Christoph Aisleitner
p. 354 left

© Markgorzynski
pp. 68-69

© Mat Honan
p. 66 left

© Matthieu Sontag
p. 146 left

© Matthijs Wetterauw
pp. 156-157

© Mauricio Mercadante
p. 236 left

© mauroguanandi
pp. 292, 294-295, 296-297, 298-299, 428 left, 430-431

© Mexrix
pp. 28-29

© Michael Clarke staff
pp. 284-285

© Michael Shake
pp. 60-61

© Miguel Vieira
pp. 62-63, 100 right, 150 right, 268-269

© Mike Brake
pp. 356-357

© Mike Laptev
pp. 216-217

© Mirgolth
p. 14 left

© Mitch Barrie
pp. 82-83

© mms_geek_chic
p. 340 right

© MONGO
pp. 384 right, 424 right

© Mwanner
p. 54 left

© NaJina McEnany
p. 416 right

© nbonzey
pp. 132-133

© Nicholas A. Tonelli
pp. 90 left, 288-289, 320-321, 438-439

© Nova
p. 154 right

© Oona Räisänen
p. 44 right

© Operation Shooting
pp. 434-435

© Pablo Alberto Salguero Quiles
pp. 76 right, 360 right, 458 right

© Pablo D. Flores
p. 380 upper left

© PatagoniaArgentina
p. 118 left

© Paul A. Mistretta, USDA Forest Service
p. 148 left

© PAUL ATKINSON
pp. 276-277

© Pi-Lens
pp. 108-109

© Piedmont NWR for the US Fish and Wildlife Service
p. 148 right

© plantsforpermaculture
p. 204 right

© Prazak
p. 342 upper left

© Pymouss
p. 436 left

© R1CH
p. 432 left

© R.L.Hausdorf
pp. 112-113

© Rachel_thecat
p. 36 right

© Rafal fabrykiewicz
pp. 226-227

© Raffi Kojian
pp. 300 upper left and right

© Ricardo Cordeiro
pp. 18-19

© Richard Webb
p. 120 right

© riekephotos
pp. 198-199

© Riverbanks Outdoor Store
p. 162 right

© Roban Kramer
pp. 326-327

© Robert H. Mohlenbrock for USDA-NRCS PLANTS Database
p. 134 right

© Roland Tanglao
p. 94 left

© Rotational
pp. 248 upper left and right, 466 right

© S. Saramart
pp. 256-257

© Sally Scott
pp. 72-73

© Sanchezn
p. 476 left

© Sannse
p. 230 left

© Shizhao
p. 440 right

© Shutterschok
pp. 232-233

© Silversypher
pp. 52-53

© spirit of america
pp. 102-103

© Stacey Meadwell
p. 114 right

© Stan Shebs
p. 282 left

© Steffen Foerster
pp. 32-33, 34-35

© Sten Porse
p. 336 right

© Steve Wood
pp. 78-79

© StevenRussellSmithPhotos
pp. 280-281

© Sue Waters
pp. 176-177, 200 left

© Svein Harkestad
p. 214 right

© TANAKA Juuyoh
pp. 330-331, 332-333, 334-335

© Tania Sohlman
pp. 46-47

© TeunSpaans
pp. 228 right, 324 right

© Thomas & Dianne Jones
p. 100 left

© Tiago Fioreze
p. 462 left

© Tilo Podner
p. 80 right

© Tusharkoley
pp. 240-241

© Urban
p. 84 right

© Urosr
pp. 86-87

© U.S. Fish and Wildlife Service
p. 64 left

© US Forest Service Dorena Genetic Resource Center
p. 146 right

© USDA Forest Service – Region 2 – Rocky Mountain Region Archive, Bugwood.org
p. 58 left

© United States Department of Agriculture
p. 228 upper left

© Vassil
p. 258 right

© Villiers Steyn
pp. 250-251

© Visitor7
p. 238 left

© Vitoriano Junior
pp. 170-171

© Vulcano
p. 360 left

© Walid Nohra
pp. 190-191

© Walter Siegmund
pp. 30, 70 left, 110 right, 130 left, 154 left, 196 left, 238 right, 416 left, 420-421

© Webgoddess
p. 184 right

© Wendy Cutler
pp. 166-167, 178-179, 264-265, 310-311, 312-313, 316 left, 322-323, 406 right, 408 right, 410-411

© Wibowo Djatmiko
pp. 218, 414

© Wildnerdpix
pp. 88-89, 96-97, 186-187

© Willow
p. 262 left

© Woodlot
p. 150 left

© Worldstar22
p. 336 lower left

© Wouter Hagens
pp. 380 lower left, 472 lower left

© Wyatt Berka
p. 58 left

© Xico Putini
pp. 172-173

© Yann
p. 412 left

© YuryZap
pp. 442-443

RELATED WEBSITES

Agroforestry Tree Database
www.worldagroforestry.org/Sea/Products/AFDbases/AF/index.asp

Archtoolbox
archtoolbox.com/materials-systems/wood-plastic-composites.html

Boreal Forest
www.borealforest.org/index.php

Center for Wood Anatomy Research, USDA Forest Service, Forest Products Laboratory
www.fpl.fs.fed.us

Convention on International Trade in Endangered Species of Wild Fauna and Flora
www.cites.org

Guía de las principales maderas y de su secado
www.guiadelasmaderas.com

History of Forestry
www.lib.ncsu.edu/specialcollections/forestry/index.html

Hobbit House
www.hobbithouseinc.com/personal/woodpics/index.htm

Hortipedia, the Garden Info Portal
http://en.hortipedia.com/wiki/Category:Trees

International Wood Collectors Society
www.woodcollectors.org

Naturally Wood
www.naturallywood.com/wood-products

Northern Arizona University School of Forestry
www.for.nau.edu/courses/for212/taxonomy.htm

NZ Wood
www.nzwood.co.nz

Oxford Plant Systematics
herbaria.plants.ox.ac.uk/herbaria_pages/xylarium.html

Plants Database, USDA Natural Resources Conservation Service
plants.usda.gov/java

Society of Wood Science and Technology
www.swst.org/teach/teach2/properties2.pdf

Talla Madera
www.tallamadera.com

Tervuren Xylarium Wood Database
www.metafro.be/xylarium

The Australian Timber Database
www.timber.net.au

The IUCN Red List of Threatened Species
www.iucnredlist.org

The reference resource for timbers of the world
sites.google.com/a/woodbook.co.uk/www/home2

The TAXA Wood Knowledge Base
www.woodsoftheworld.org

The Wood Database
www.wood-database.com

The Wood Explorer
www.thewoodexplorer.com

TIMSPEC Importers of specialized timbers
www.timspec.co.nz/Learning-Centre/Cuts-of-Timber-8170.htm

Tree ring research on conifers in the Alps
christian.rolland.free.fr/index.htm

Veneer selector
www.veneerselector.com

Wikimedia Commons
commons.wikimedia.org/wiki/Main_Page

Wikipedia
en.wikipedia.org/wiki/Main_Page

Wikispecies
species.wikimedia.org/wiki/Main_Page

Wood anatomy of Central European species
www.woodanatomy.ch

Woodworkers Source
www.woodworkerssource.com

Xiloteca Manuel Soler
www.xiloteca.com/publicaciones.asp